The Aesthetics of the Undersea

Among global environments, the undersea is unique in the challenges it poses – and the opportunities it affords – for sensation, perception, inquiry, and fantasy. *The Aesthetics of the Undersea* draws case studies in such potencies from the subaqueous imaginings of Western culture, and from the undersea realities that have inspired them. The chapters explore aesthetic engagements with underwater worlds, and sustain a concern with submarine "sense," in several meanings of that word: when submerged, faculties and fantasies transform, confronting human subjects with their limitations while enlarging the apparent scope of possibility and invention. Terrestrially-established categories and contours shift, metamorphose, or fail altogether to apply. As ocean health acquires an increasing share of the global environmental imaginary, the histories of submarine sense manifest ever-greater importance, and offer resources for documentation as well as creativity.

The chapters deal with the sensory, material, and formal provocations of the underwater environment, and consider the consequences of such provocations for aesthetic and epistemological paradigms. Contributors, who hail from the United States, United Kingdom and Australia, include scholars of literature, art, new media, music and history. Cases studies range from baroque and rococo fantasies to the gothic, surrealism, modernism, and contemporary installation art. By juxtaposing early modern and Enlightenment contexts with matters of more recent – and indeed contemporary – importance, *The Aesthetics of the Undersea* establishes crucial relations among temporally remote entities, which will resonate across the environmental humanities.

Margaret Cohen is Andrew B. Hammond Professor of French Language, Literature, and Civilization at Stanford University, where she teaches in the Departments of Comparative Literature and English.

Killian Quigley is Postdoctoral Research Fellow at the Sydney Environment Institute at the University of Sydney.

Routledge Environmental Humanities

Series editors: Paul Warde (University of Cambridge, UK) and Libby Robin (Australian National University)

The *Routledge Environmental Humanities* series is an original and inspiring venture recognising that today's world agricultural and water crises, ocean pollution and resource depletion, global warming from greenhouse gases, urban sprawl, overpopulation, food insecurity and environmental justice are all *crises of culture*.

The reality of understanding and finding adaptive solutions to our present and future environmental challenges has shifted the epicenter of environmental studies away from an exclusively scientific and technological framework to one that depends on the human-focused disciplines and ideas of the humanities and allied social sciences.

We thus welcome book proposals from all humanities and social sciences disciplines for an inclusive and interdisciplinary series. We favour manuscripts aimed at an international readership and written in a lively and accessible style. The readership comprises scholars and students from the humanities and social sciences and thoughtful readers concerned about the human dimensions of environmental change.

The Aesthetics of the Undersea

Edited by Margaret Cohen and Killian Quigley

Routledge
Taylor & Francis Group
LONDON AND NEW YORK

earthscan
from Routledge

First published 2019
by Routledge
2 Park Square, Milton Park, Abingdon, Oxon OX14 4RN

and by Routledge
711 Third Avenue, New York, NY 10017

Routledge is an imprint of the Taylor & Francis Group, an informa business

British Library Cataloguing-in-Publication Data
A catalogue record for this book is available from the British Library

Library of Congress Cataloging-in-Publication Data
Names: Cohen, Margaret, 1958- editor. |
Quigley, Killian Colm, editor.
Title: The aesthetics of the undersea / edited
by Margaret Cohen and Killian Quigley.
Description: Abingdon, Oxon ; New York, NY : Routledge, 2019. |
Series: Routledge environmental humanities |
Includes bibliographical references and index.
Identifiers: LCCN 2018030301 (print) |
LCCN 2018050437 (ebook) | ISBN 9780429444203 (eBook) |
ISBN 9780367001582 (hardback) | ISBN 9780429444203 (ebk)
Subjects: LCSH: Ocean--Miscellanea. | Nature (Aesthetics) |
Environment (Aesthetics) | Ocean and civilization.
Classification: LCC BH301.S4 (ebook) | LCC BH301.S4 A37 2019 (print)
| DDC 809/.9332162--dc23
LC record available at https://lccn.loc.gov/2018030301

ISBN: 978-0-367-00158-2 (hbk)
ISBN: 978-0-429-44420-3 (ebk)

Typeset in Bembo
by Integra Software Services Pvt. Ltd.

Contents

Figures

Contributors

Samuel Baker teaches in the English Department at the University of Texas, USA. He has published a book, *Written on the Water: British Romanticism and the Maritime Empire of Culture* (The University of Virginia Press, 2010), and essays on authors including Ann Radcliffe, William Wordsworth, Walter Scott, and Mary Shelley. His current research interests include media theory and the history of the gothic.

Margaret Cohen holds the Andrew B. Hammond Chair of French Language, Literature, and Civilization at Stanford University, USA, where she teaches in the Departments of Comparative Literature and English. Her books include *The Novel and the Sea* (Princeton University Press, 2010), *The Sentimental Education of the Novel* (Princeton University Press, 1999) and *Profane Illumination: Walter Benjamin and the Paris of Surrealist Revolution* (University of California Press, 1993). She is currently completing a book on the aesthetics of underwater film and is editing the six-volume *A Cultural History of the Sea* forthcoming with Bloomsbury Press.

Natalie Deam is a PhD Candidate in the Department of French and Italian at Stanford University, USA. Her dissertation, *The Fantastic Natural and the Evolutionary Imagination in Nineteenth-Century France*, examines the ways that conflicting evolutionary theories provoked a subversive, fantastic imaginary of nature that appears across diverse literary genres in France and questions modern understandings of race, gender, and sexuality.

Ann Elias is Associate Professor in the Department of Art History at the University of Sydney, Australia, and a Key Researcher with the Sydney Environment Institute. Research interests include: camouflage as a military, social and aesthetic phenomenon; flowers and their cultural history; coral reef imagery and the underwater. Her latest book, about early underwater photography, is *Coral Empire,* forthcoming with Duke University Press.

Prudence Gibson is a Postdoctoral Research Fellow at UNSW Art & Design, Australia. Along with a long track record of art writing, she is also author of the following: *The Rapture of Death* (Boccalatte Books, 2010); *Janet Laurence: The Pharmacy of Plants* (UNSW Press, 2015); *The Plant*

Contract (Brill-Rodopi, 2018); *Aesthetics After Finitude* (Re.Press, 2016); *Covert Plants* (Punctum Books, 2018).

Robin Jarvis is Professor of English Literature at the University of the West of England, UK. His publications include *Romantic Writing and Pedestrian Travel* (Macmillan, 1997), *The Romantic Period: The Intellectual and Cultural Context of English Literature, 1789-1830* (Pearson Longman, 2004), and *Romantic Readers and Transatlantic Travel* (Ashgate, 2012), and numerous articles and book chapters in the fields of Romanticism and travel writing.

Jonathan Lamb has taught at the universities of Auckland, Princeton and Vanderbilt, where he holds the Andrew W. Mellon Chair of the Humanities. Last year he published *Scurvy: The Disease of Discovery* (Princeton University Press, 2016) which won an honourable mention at this year's Gottschalk prize-giving at the American Society for Eighteenth-Century Studies. Currently he is editing a collection of essays in the Bloomsbury series *A Cultural History of the Sea* and co-editing a collection of keywords on reenactment. He is researching a monograph on the aesthetics of the ellipse.

Janet Laurence is a Sydney-based Australian artist who exhibits nationally and internationally. Her practice examines our physical, cultural and conflicting relationship to the natural world. She creates immersive environments that navigate the interconnections between organic elements and systems of nature. Within the recognized threat to so much of the *life world* she explores what it might mean to heal, albeit metaphorically, the natural environment, fusing this with a sense of communal loss and search for connection with powerful life-forces. Her work is included in museum, university and corporate and private collections as well as within architectural and landscaped public places. www.janetlaurence.com.

Iain McCalman is Research Professor in the Department of History at the University of Sydney, Australia and co-Director of the Sydney Environment Institute. His award-winning book, *The Reef – A Passionate History, from Captain Cook to Climate Change* (2014, 2016), was published by Penguin in Australia and by Farrar, Straus and Giroux/Scientific American in the USA. He is a Fellow of four Learned Academies and is a former President of the Australian Academy of the Humanities. He was Director of the Humanities Research Centre, ANU, from 1995-2002.

Sean Theodora O'Hanlan is a PhD candidate in the Art & Art History Department at Stanford University, USA. She is the 2017–2019 Leonard A. Lauder predoctoral fellow at the Leonard A. Lauder Research Center for Modern Art at the Metropolitan Museum of Art, where she is completing her dissertation, *André Breton and the Modern Art of Collecting*.

Killian Quigley is a Postdoctoral Research Fellow at the Sydney Environment Institute, University of Sydney, Australia. His work is available or forthcoming in *Eighteenth-Century Life, The Eighteenth Century: Theory and Interpretation,*

Ireland's Imperial Connections, 1775-1947 (Palgrave), and *A Cultural History of the Sea in the Age of Enlightenment* (Bloomsbury). He is preparing a monograph on submarine aesthetics and poetics in eighteenth-century Europe.

Luis Rodríguez Rincón of the Department of Comparative Literature, Stanford University, USA is a student of the early modern period. A fascination with how the ocean was represented and understood in the wake of 1492 has led him to investigate the ways grottos, nymphs, and tritons were used to represent a terraqueous world in the poetry and prose of Italy, Spain, Portugal, France, and England over the course of the long sixteenth century.

Hanna Roman is Assistant Professor of French at Dickinson College, USA. She studies the discourses of science in Enlightenment France. Her book, *The Language of Nature in Buffon's 'Histoire naturelle'*, is forthcoming with Oxford University Studies in the Enlightenment in October, 2018.

Linda Williams is Associate Professor of cultural and environmental history at RMIT University, Australia, where she leads a research network in the field of environmental art and cultural studies, and has curated several international exhibitions including *Ocean Imaginaries* in 2017 at the RMIT Gallery in central Melbourne. Her recent publications in the field of the environmental humanities can be accessed at https://rmit.academia.edu/LindaWilliams.

Josh Wodak is a researcher, writer and artist whose work critically engages with cultural and ethical entanglements between ecological engineering and conservation biology as means to mitigate species extinction and biodiversity loss in the Anthropocene. He holds a BA (Honours) in Anthropology (The University of Sydney, 2002) and a PhD in Interdisciplinary Cross-Cultural Research (Australian National University, 2011). He is currently Lecturer, UNSW Art & Design, Australia, a Chief Investigator on the Australian Research Council Discovery Project *Understanding Australia in The Age of Humans: Localising the Anthropocene* and a member of the Andrew Mellon Australia-Pacific Observatory in Environmental Humanities, Sydney Environment Institute, University of Sydney.

Acknowledgments

This volume springs from several projects under the auspices of "Visualizing the Oceans," a collaborative grant, Margaret Cohen Principal Investigator, supported by the Stanford Arts Institute and the Stanford Humanities Center. The Sydney Environment Institute and Vanderbilt University co-funded the projects, including the 2015 conference "The Underwater Realm," co-sponsored by Iain McCalman and Jonathan Lamb. The Sydney Environment Institute launched preparation of the volume, and Stanford University provided support for images.

Diving into the Wreck

First having read the book of myths,
and loaded the camera,
and checked the edge of the knife-blade,
I put on
the body-armor of black rubber
the absurd flippers
the grave and awkward mask.
I am having to do this
not like Cousteau with his
assiduous team
aboard the sun-flooded schooner
but here alone.

There is a ladder.
The ladder is always there
hanging innocently
close to the side of the schooner.
We know what it is for,
we who have used it.
Otherwise
it's a piece of maritime floss
some sundry equipment.

I go down.
Rung after rung and still
the oxygen immerses me
the blue light
the clear atoms
of our human air.
I go down.
My flippers cripple me,
I crawl like an insect down the ladder

and there is no one
to tell me when the ocean
will begin.

First the air is blue and then
it is bluer and then green and then
black I am blacking out and yet
my mask is powerful
it pumps my blood with power
the sea is another story
the sea is not a question of power
I have to learn alone
to turn my body without force
in the deep element.

And now: it is easy to forget
what I came for
among so many who have always
lived here
swaying their crenellated fans
between the reefs
and besides
you breathe differently down here.

I came to explore the wreck.
The words are purposes.
The words are maps.
I came to see the damage that was done
and the treasures that prevail.
I stroke the beam of my lamp
slowly along the flank
of something more permanent
than fish or weed

the thing I came for:
the wreck and not the story of the wreck
the thing itself and not the myth
the drowned face always staring
toward the sun
the evidence of damage
worn by salt and sway into this threadbare beauty
the ribs of the disaster
curving their assertion
among the tentative haunters.

This is the place.
And I am here, the mermaid whose dark hair
streams black, the merman in his armored body
We circle silently
about the wreck
we dive into the hold.
I am she: I am he

whose drowned face sleeps with open eyes
whose breasts still bear the stress
whose silver, copper, vermeil cargo lies
obscurely inside barrels
half-wedged and left to rot
we are the half-destroyed instruments
that once held to a course
the water-eaten log
the fouled compass

We are, I am, you are
by cowardice or courage
the one who find our way
back to this scene
carrying a knife, a camera
a book of myths
in which
our names do not appear.

Adrienne Rich

Introduction

Submarine aesthetics

Margaret Cohen and Killian Quigley

The environment beneath the waves is so alien from our terrestrial reality that it is like another world – and yet we reach it simply by stepping off the edge of the shore. Poet Adrienne Rich used its otherness to confront the most intractable aspects of the human condition in "Diving Into the Wreck," the poem that gave the title to her 1973 collection. "[H]aving read the book of myths, /and loaded the camera," the poet dives down on a quest to find "the wreck and not the story of the wreck/the thing itself and not the myth."[1] Yet as she descends, the bizarre conditions of the undersea warp the clarity of her perceptions. She is perturbed by gradual darkening and loss of color as she dives away from the light of the sun, and by increasing pressure on her body: "the air is blue and then/it is bluer and then green and then/black I am blacking out." She is beguiled and distracted by strange submarine flora and fauna and "forget[s]/ what I came for/among so many who have always/lived here/swaying their crenellated fans." In Rich's sensuous evocation, "the deep element" beckons for a dive into revelation – and yet these conditions simultaneously threaten the diver's ability to realize the mission that called her down.

Rich's poem exemplifies the generative potencies of submarine environments, for aesthetics and for the concerns that are its orbit, such as philosophy, culture, and critical analysis. *The Aesthetics of the Undersea* draws case studies in such potencies from the subaqueous imaginings of Western modernity, and from the undersea realities that have inspired them. Rich's poet has read the book of myths, but crucially, the realm she encounters is not mythological, and awareness of its specificity is integral to her quest. An aesthetic history of the submarine acknowledges contingent cultural inheritances and essential physiological realities simultaneously. Thinking the submarine in relation to aesthetics and the physiology of the senses emphasizes the materialities of the undersea, as well as its material impacts upon immersed bodies. At the same time, our approach treats as crucial extra-physiological routes to sensation, such as memory, natural history, and repertoires of fantasy.

"[A]nd besides/you breathe differently down here": when submerged in an element so different from air, faculties and fantasies transform. Terrestrially-established categories shift, metamorphose, or fail altogether to apply. Human bodies confront stark physiological limitations, starting with our lack of a way to extract oxygen from water. Without the buffer of a layer of air, human vision does not work properly beneath the sea. Orientation struggles where molecules eight hundred times denser than air magnify objects by about 33 percent, and where the terrestrial sense of gravity transforms. Prosthetic instruments – from goggles, glass-bottomed boats, and diving regulators to submersibles and remotely-operated vehicles – often supplement or supplant the senses underwater, troubling and enriching the human status of the undersea-goer with qualities of machine, or android. At the same time, immersion opens new pathways for sensation, perception, inquiry, and imagination. Often, subaquatic boundaries to cognition and convention are portals to new forms, lives, and epistemologies. The undersea has long invited art and literature, as well as science, to detect and conceive vast new realms of possibility.

The specificity of the submarine's aesthetic atmosphere helps rumple firm separations between aerosphere, surface, and undersea. Such rumpling involves the challenge of adapting terrestrially-inflected terms, frames, and indeed senses across these zones. Since human vision, for example, is altered, diminished, and eventually voided in the depths, Western modernity's famous ocularcentrism begins to have a distinctly difficult, if occasionally distinctly interesting, time. One representative aesthetic difficulty involves the sublime, which since the eighteenth century has been identified with pleasurable topside scenes, such as Joseph Addison's "wide expanse of water"[2] and Immanuel Kant's sea-storms.[3] Sublimity has trouble underwater, though, largely because the aesthetic subject, or spectator no longer has the luxury of disinterested and securely distanced contemplation, which Edmund Burke and others identified as crucial to its proper functioning.[4] The sublime's subaqueous awkwardness figures a crowd of associated problems, pertaining to subject, object, sentiment, prospect, scene, and so forth. And this is only one instance: generally speaking, inherited vocabularies and conventions in Western aesthetic history are put out of whack by underwater realms.

Currents in undersea aesthetics have formed in time, and in proximity to culture. In addition to attending to submarine materialities, *The Aesthetics of the Undersea* pays heed to the situated historicity of submarine experience. In the first stanza of "Diving Into the Wreck," Rich's poet describes preparing for her exploration: "I am having to do this/not like Cousteau with his/assiduous team/. . . but here alone." At the time Rich wrote "Diving into the Wreck,"the TV series *The Undersea World of Jacques Cousteau* brought undersea vistas, and the work of undersea exploration, into homes around the globe. Land-lubbing audiences were captivated by the kinds of granular technical detail – such as the effects of depth on light perception and divers' air supplies – and spectacular subaqueous prospects that infuse Rich's poem. Rich's marvelous alien landscape is one of many ways of imagining the submarine realm, but it is unthinkable without the scuba and underwater filmmaking techniques pioneered by Cousteau and others.

So this poet's views of swaying crenellated fans are as much of their time as the TV show. In the 1970s, those fans could seem "always/[to have] lived here." Permanence and timelessness are recurring tropes in undersea poetics, but their evocative powers are today under insuperable pressure from climate change and other phenomena. Such pressures are inscribed upon this book. In 2014, we first conceived of the collection during a research expedition to the Great Barrier Reef. In the short time since, some of the vibrant coral formations we viewed, notably around Heron Island, have been destroyed by catastrophic bleaching.

The undersea contains, reflects, and expresses histories. Likewise, modes of submarine aesthetics are and have always been situated at complex historical and cultural nexuses. The technologies of Western modernity play a central role in transforming underwater access across the aesthetics examined in this collection, from the early modern era, when Europeans began charting the high seas of the globe, to the deep-sea submersibles of the later twentieth century. A few of the stories this collection tells concern the impacts of shifts in underwater access for how submarine environments could be imagined and represented. For instance, Killian Quigley's chapter considers the rococo's flourishing in early eighteenth-century Europe, and the proliferation of liquid, shelly, coralline, and otherwise oceanic forms and colors that attended and contributed to it. Those forms and colors were fished from the sea, and from fancy. Coincident with marinal finery was a submarine epistemology in flux: the rococo owed its imaginings partly to the cabinets of curiosity that had long featured shells and coral skeletons prominently, partly to early attempts at sea-going empiricism and the odd materials and lives they encountered, and partly to land-borne aesthetic principles from painting, the plastic arts, dance, and elsewhere. This effort to imagine the sea into land is also evident in the eighteenth-century natural histories of the sea surveyed by Hanna Roman, who emphasizes philosophers' interest in using the undersea as an ultimate challenge for the Enlightenment project to make the invisible visible. The terrestrial orientation was a far cry from what Robin Jarvis calls the romantic era's "hydromania," as middle- and upper-class populations began swimming in droves, intoxicated with "swimming sense," as they immersed their bodies in the sea.

While romantic swimmers did not have technologies yet to peer or linger under the water, their willingness to brave its challenges with their bodies was essential to the penetration of Western societies into the sea. It encouraged human immersion, via swimming and diving, which has at crucial junctures been othered as irrational, uncivilized, and animal. In *The Lure of the Sea* (1988), Alain Corbin influentially described Europe's re-visioning of the seashore in the late eighteenth and early nineteenth centuries, as a space of hygiene and pleasure, and not danger. It bears noting that this shift, from hydrophobia to Jarvis's hydromania, contrasts with the practices of non-Western marine-oriented cultures of the same era. Europeans knowledgeable in the ways of the sea, like members of James Cook's voyages into the Pacific during the 1770s, were amazed by the adroit feats of South Pacific islanders in the water. When Christianity and sugar plantations came to Hawaii in the nineteenth century, missionaries framed swimming, and other Hawaiian water sports like surfing, as sinful encouragements to naked co-mingling of the sexes. For the planters, these sports were irritating disruptions to the schedule of plantation labor. If *The Aesthetics of the Undersea* attends to the immersive history of Western modernity, it is also a witness to Western imperialism: beneath the waves as across the globe, in aesthetics as in capitalism and conquest.

As technologies advanced enabling underwater breathing, from the closed-helmet diving suit of the nineteenth century to the scuba gear of the twentieth, humans came to see the reality of the depths. Prosthetic instruments enabled by the industrial revolution played a vital role in extending human perception into the sea from the late nineteenth, and above all from the twentieth centuries. The prominence of technologies in underwater access threads through our study, as does the historical state of scientific conceptions of the seas. Materialist analysis of creative expression has learned from Pierre Bourdieu to take account of the formative impact of the cultural field, along with the workings of society, economics and politics. A materialist analysis of environmental aesthetics – notably one engaging such a forbidding realm as the undersea – teaches the need to take into account the specificity of the environment being represented as well as the science and technology of accessing it. In the case of the undersea, theology too has shaped its conception across the span of our collection. In the early modern era, the oceans were still scaled to Biblical history. By the turn of the nineteenth century, this theological measure was coming into conflict with the scale of natural, evolutionary history. The undersea as the domain of slow evolutionary time has been complicated once more in recent years by recognition of the pressures of the Anthropocene.

As ocean health occupies an increasing share of the global environmental imaginary, the significance of submarine sense, and of its histories, manifests ever greater importance. Coral is a key index of the seas' condition, and a key preoccupation of our volume: not only in terms of transformations in its perception over time, but of transformations in coral structures in the world. The Great Barrier Reef and other threatened undersea habitats have been the subjects of path-breaking work in recent documentary filmmaking, from *The Blue Planet* (2001) and its sequel to *Chasing Coral* (2017). *The Aesthetics of the Undersea* brings historical and aesthetic contexts to bear on representations like these.

Our work not only makes better-informed readers and viewers of undersea artefacts; it deepens and complicates the archives available to documentarians and to all who strive to connect publics to ocean health. The Enlightenment's sense of the undersea as the frontier of the visible has perhaps diminished, but it has not abated. Representations of its dark depths can play an invaluable role in convincing publics of the responsibilities corporations bear for devastating the marine environment. To cite just one recent instance, the photos of undersea plumes gushing on the seafloor out of the broken pipe of the Macondo well drilled for British Petroleum in the depths of Gulf of Mexico sensitized global audiences to the reaches of the 2010 disaster invisible from surface and shores.

Further, we encourage documentarians to take note of the aesthetics that shape their methods of appealing to audiences' emotions. "[I]t was like trying to catalog a dream" recalls Elizabeth Kolbert of the profusion of undersea life on the pristine coral reefs around One Tree Island, a remote cay on the Great Barrier Reef.[5] The portrayal of the ocean depths as a dream dates to an avant-garde lineage figuring unconscious processes and intimating a new surreal realm, one obliquely accessible through weird images of the submarine environment. As well as a dream world, this avant-garde lineage likened the aquatic realm to a drug trip, an aesthetic Kolbert cleverly ironizes in her chapter title, "Dropping Acid," which refers to ocean acidification. While Kolbert's rhetoric is arresting, is her irony sufficient safeguard against modernist habits of turning the underwater environment into a figure for an exotic elsewhere? With her descriptive rhetoric, Kolbert risks mobilizing what might be called an "environmental orientalism" that encourages audiences to luxuriate in faraway places. One Tree Island faces imminent threats from rising sea temperatures and ocean acidification. Energizing audiences while conveying the realities of climate change is of course an immense challenge facing Kolbert and all those who popularize the dire findings of science. Our case studies show how the humanities have a vital role to play in better understanding the documentation of underwater states, as well as their fantasies.

In emphasizing specificity – of time, of place, of aesthetics – our case studies challenge a weighty strain in Western imaginations of the oceans as unbounded and unlocated. This strain runs through religion and secular romanticism, and holds power even now. While the undersea is thinkable as a concept – this book's title relies upon it – it is not an abstraction. It is an area that spans the globe and contains extremes, from the shallow zones of the intertidal to abyssal depths, and from the bathwaters of the tropics to the polar oceans. Moreover, these places and depths contain both naturally-occurring flora and fauna and invasive species, as well as human traces, impacts, and remains. When this book's analyses contain the undersea within a single aesthetic frame, they nonetheless stress connections between submarine materiality and aesthetics, and so retain and recover the specificity of underwater place. We thus hope to provide a corrective to tendencies that theorize aquatic aesthetics as the other of landed fixedness, and that prioritize dissolution into a Deleuzian flow.

Further, it would be intellectually, ethically, and politically disastrous to jettison notions of locality and regionalism from the waters. This volume's aesthetic instances are situated, and demonstrate the potential for aesthetic, and more broadly humanistic, inquiry to prove useful for all discourses that address the peculiarities of submarine place. As recent work in undersea geography makes distinctly clear, these knotty and provocative matters are materially and legally significant. Radical differences in underwater environments transfer themselves to radical differences in feeling, description, and definition – and these incongruities call into question the submarine portability of structures like the European Landscape Convention.[6] Legal questions have long haunted marine artefacts, as they have the zonation of who owns the seas. The wreckage of the *Titanic* provided stunning gothic images for James Cameron in the film *Titanic* (1997), and yet who owns the rights to the wreck?[7] The question was the subject of litigation, given the value of the remains, and similarly vexed questions apply to other shipwrecks littering the world's oceans. The (fruitless) collaboration of China, Australia, and Malaysia to locate the black box from MH370 lost in 2014 indicates the specific political and social interests that apply even beneath the sea.

The Aesthetics of the Undersea enters a scholarly milieu that has, for the past few decades, been increasingly engaged with environment, ecology, and related concepts, and increasingly activated by ecosystem degradation, environmental injustice, and more-than-human ontologies. Work in Western environmental aesthetics has, likewise, been thinking and moving around and past its inheritances, notably the long-standing dichotomy between humans (and cultures) and nature. Participation, immersion, entanglement, reciprocity, intimacy, and embodiment are just some of the energies that have recently gained currency among theorists keen to operate beyond and between busy subjects and objective – often inert – scenery. For many reasons enumerated in this introduction, and these chapters, the submarine would seem to be an ideal atmosphere for a more progressive environmental aesthetics. Beneath the waves, human senses are materially implicated in their surroundings to an extent that defies attempts to ignore or obscure their contingency. Technologies broaden the scope of detection, but in so doing affirm the mediateness of underwater impression. As some of those few environmental critics who have attended to the underwater environment note, the undersea's consequences for human subjectivity and human subjects – which are variously and differentially challenged, foreclosed, and otherwise altered in underwater worlds – have the potential to encourage new (post)humanisms.

Blue cultural studies, ocean studies, the blue humanities, the maritime turn, "thalassology": these are a few names for recent seaward movements in humanities research. This collection contributes to this paradigm as well, and especially to the efforts of those who apply themselves to the undersea. A social impetus renewed the marine and maritime humanities in the 1990s, prompting scholars to look at seas, as well as lands, as theaters of social relations. Strangely, the specificity of oceanic environments – and particularly of their subsurface ranges – was sometimes lost in this creative social turn. But thanks to a more general environmental awareness in the humanities over the past ten years, as well as growing attention to the seas' vulnerability today, scholarly attention has recently been more actively submerged. *The Aesthetics of the Undersea* helps introduce underwater environments, in their granularity and historicity, as constitutive concerns of the blue humanities. Given the fact that these depths were accessed most extensively with industrial technologies, it is to be expected that the incorporation of the undersea in marine and maritime studies has tended to neglect earlier eras. In contrast, this book engages meaningfully with early modern and eighteenth-century contexts. Another consequence of the importance of science, technology and innovation in accessing the undersea has been relative scholarly neglect of the aesthetic meanings of submarine materiality. Those meanings, and the methodologies they enable, are at the core of the conversations our collection aims to begin, and to amplify.

One of the claims this book makes is that the undersea *is* a vital atmosphere for revising and reorienting aesthetics, and for rejigging humanistic thought. Operating across and among literary criticism, philosophy, histories of art and science, and creative practice, our collection suggests the extent to which the marine disrupts normative boundaries between epistemological, methodological, and disciplinary domains. Subaqueous phenomena can appear to frustrate humanism's fundamental terms: much as human-submarine relations vary enormously across place, culture, and time, the undersea is basically not anthropic. At certain junctures, this has perhaps been exceptionally true of the range of cultural traditions considered in this volume. *The Aesthetics of the Undersea* does not exoticize its subjects, but it does draw many of its essential energies from the submarine's slippages and strange workings, and its tendency to disobey the expectations carried to it by Western tradition. The entanglements of aesthetic convention and underwater world described in our chapters do not encompass the undersea. Ultimately, the work they perform is as much about suggesting the limits of those entanglements, and detecting the marvelous aesthetic possibilities that might lie beyond them.

We have sequenced the chapters of *The Aesthetics of the Undersea* chronologically, from the baroque grotto that opens the volume to the twenty-first century hospital for coral care that closes it. We do so to invite readers to trace crucial historical shifts in technologies of underwater access, fantasy and representation, yet any sense of teleological progress is complicated by our cases' many interrelations, several of which span large temporal, cultural, and formal ranges. Case studies include the baroque, the rococo, Romanticism, modernism, surrealism, the gothic, pop music, and installation art, along with documentary works in illustration, photography, and film. These submarine repositories, it should be stressed, are not surveys, and are not exhaustive. For example, to take up only the later nineteenth century, we do not include chapters on the popular exhibition form of the aquarium, Impressionism's sea fantasies in art and music, or even science fiction, like the foundational work of undersea adventure by Jules Verne, *Twenty Thousand Leagues Under the Sea (Vingt mille lieues sous les mers)* (1870). The volume is an invitation to attend to the expanse of the undersea imagination, an expanse far too extensive for its covers. We are anticipating fellow-collections to broaden the scope of submarine inquiry: we look forward, for example, to the publication of *Underwater Worlds: Submerged Visions in Science and Modern Culture* (Cambridge Scholars Press, forthcoming 2019), edited by William Abberley, which originated at a conference we had the pleasure of attending at TORCH in 2015. As Abberley explains, the collection limns how "representations of subaqueous environments have changed in the modern period," revealing across its essays "that the remoteness and alterity of underwater environments accentuate the contingencies and artifice of representation."[8]

Among the opportunities afforded to the imagination by the undersea in our collection is its ability to mobilize the uncanny. The switch between familiar and unfamiliar happens reciprocally – both when underwater elements are brought to land and when human technologies are submerged beneath the sea. Such switching mobilizes a panoply of moods, from the stately mastery of baroque grottoes and rococo whimsy to the modernist tragedy of the inhuman spectacle of nature. A powerful aesthetic inspiration is the "déreglement de tous les sens" – the disorientation and denormalization of all sense, senses and meaning – that immersion effects, in the words of Arthur Rimbaud, symbolist poet and submarine visionary in his "The Drunken Boat" ("Le Bateau ivre") (1871). Vision's blockage is another thread that runs throughout the volume, most notably in Enlightenment contexts. The undersea thus acquires a fascinating status as at once disrupting illumination and initiating fantasy explorations. The water's surface enters into this complex coherence as well, as a barrier to immersion and as a threshold.

The volume opens with an essay on the baroque grotto, a site of leisure and fantasy in the early modern era, creating a rustic ambiance at odds with renaissance innovations of classical beauty. Focusing on marine-themed grottos, Luis Rodríguez Rincón reveals that this rustic allure was no throwback, but rather a baroque convergence of timely innovations fused with mystic forms. In antiquity, water nymphs lived in sacred grottos – but the renaissance constructs used sophisticated hydraulic engineering to bathe the grotto in moisture and flow. For Neo-Platonists, water was dangerously pleasurable, immersed in sensuousness and earthly delights. The baroque grotto, in contrast, used water to revel in human power over physis, a power also evident in contemporary maritime technologies. Further, the grottos provided accurate, empirical depictions of marine life indebted to nascent science and Europeans' expansion over the seas. The grotto was a machine of submarine fantasy, where Neptune on his throne was no longer a furious god marking the limits of human domain but rather a proponent of seaward-looking humanism.

Baroque spectacle gives way to rococo delicacy in Killian Quigley's investigation of a recurring trope in undersea aesthetics: the resort to the language, contours, and materials of the plastic arts to describe and represent the submarine. In the early eighteenth century, the rococo expressed a pronounced "marinal sensibility," sublimating the subaqueous into porcelain, stucco, terracotta, boiserie, woodwork, and much else. Rococo exuberance has frequently been charged with superfluity and irrationality; Quigley contends that, to the contrary, decoration represents a genuinely "epistemologically productive" mode for mediating matters, spaces, and lives that did not conform to conventional categories and distinctions, as among animal, vegetable, and mineral.

Oceanic idiosyncrasies like these were the adversaries of Enlightenment French geography as described by Hanna Roman. Paramount was the imperative to render the invisible visible, and the inscrutable legible. For the authors of the monumental *Encyclopedia* (*Encyclopédie*), seawater was so much obfuscation, a mystifying layer keeping submerged land from view. That land was essentially continuous with *terra firma*, and the job of the *philosophe* was to reason "around the ocean," or through it, to access its substratum. The *je ne sais quoi* expresses a countervailing – and perhaps significantly submarine – sensibility.

Inexplicability and the reactions it provokes, for twenty-first century subjects as for seventeenth- and eighteenth-century philosophers, preoccupy Jonathan Lamb's study of color, submarine life, and Western theories of sensation and cognition. The degradation of the Great Barrier Reef involves a critical loss of chromatic life. The ideologies that refuse to take steps to protect that life reflect the legacies of the discreditation of sensory knowledge. Cartesian and Lockean abstraction fomented a deep mistrust of not only sensation, but of the body and its relationship to the environment. In view of "privations" – of color, referent, understanding, and so on – it is possible to indulge in abstract mathematical certainty, or to recognize positively the sign of the not-yet-known. Colorism, via Hobbes, Hume, Ruskin, and NASA's Portable Remote Imaging Spectrometer, suggests the latter response.

In Robin Jarvis's study of "hydromania" - or "Romantic water-lust" – bodies and language go swimming. Samuel Taylor Coleridge, Lord Byron, and John Keats – not to mention Jane Austen – experiment with immersion as a physical, poetic, social, existential, epistemological, and indeed political practice. In and from the water, it was possible not only to become singularly intimate with nature, but to render unstable the margin "between spectator and spectacle" that ran through so much of conventional aesthetics. For Romanticism, swimming became an emblem of creative dissolution, as "old certainties" fray "and a previously knowable universe becomes more fluid and indeterminate." Thus the volume marks a remarkable contrast with Roman's *philosophes*, who predated the Romantics by only a matter of decades, but regarded the sea in a dramatically different manner.

From Byron's aquatic self-actualization forward – to something a great deal less decided: Natalie Deam tracks movements simultaneously toward "the evolutionary ocean" and away from "Romanticism's easy identification" with its element. As Charles Darwin and his theories rendered the seas more lively, they also made them less comfortably human. Deam's case in point is Jules Michelet's *The Sea* (*La Mer*) (1861), a grand popular-scientific work that grappled with the weird idea of humans' marine origins, and promoted an aesthetics of the "biological sublime." Michelet trades in anthropomorphisms – often perniciously gendered ones – that do not so much establish firm ties between human beings and the ocean as untie whatever coherence either entity seemed to have as an object, or indeed as a subject position. At the same time, odd affinities do declare themselves, as between Michelet and the whales whose decimation, via hunting, inspired a marine-conservationist impulse palpable through the language of his time.

Samuel Baker's essay spans a vital stretch in the development and articulation of Western literary modernism – and of *soi-disant* Western culture. Submarine aesthetics and phenomenology were key energies for fictions, poems, and criticism that preoccupied themselves with "figures of dissociation and dissolution" in their musings on modernity. Those figures, not to mention mermen and mermaids, were here – as elsewhere – aids for fantasy-making; the undersea imaginary created spaces for metamorphoses of "terrestrial life" and terrestrial thought. The works under consideration here – from Hans Christian Andersen's "The Little Mermaid" (1846) and Matthew Arnold's "The Forsaken Merman" (1849) to T. S. Eliot's "The Love Song of J. Alfred Prufrock" (1915), as well as Eliot's theses respecting the metaphysical poets – stage subaqueous encounters between divergent frames for thought, culture, and belief. The impact from these modernist encounters will continue to echo across poetry: when Rich reaches "the place," in her dive, she becomes "we": the lyric "I," together with "the mermaid whose dark hair/streams black, the merman in his armored body."

As some literature explored the fantastic capacities of submarine figures, science and photography strove – and struggled – for subaqueous objectivity. Iain McCalman describes the challenges faced by the biologist William Saville-Kent, the first both to catalog and represent the extraordinary variety of coral on the Great Barrier Reef – without diving gear and in an era before color or underwater photography. Saville-Kent had previously mastered innovations in the aquarium, the great nineteenth-century technology for exhibiting live undersea creatures on land. In documenting corals on the Reef, he showed his mastery in nature photography, working in stringent conditions to produce informational photographs of great clarity, which, for example, could only be obtained when the coral was exposed at low tide. At the same time, in black and white photos, the distinctive color of these formations, along with their beauty, were lost. In Saville-Kent's opus, *The Great Barrier Reef of Australia: Its Products and Potentialities* (1893), reviewers appreciated access to "the thing itself," which, McCalman shows, emanated from Saville-Kent's inclusion of a range of supplementary media: black and white photos, scientific description, poetic phrase and Saville-Kent's own brilliant, finely drawn chromolithographs.

Ann Elias takes up the challenges the Barrier Reef presented to documentary representation for underwater cinema pioneer Frank Hurley, director of the notoriously racist *Pearls and Savages* (1921), whose prior experience also included photographing Shackleton's Antarctic expedition. Hurley too faced the limits of human vision and technology, as well as the way the Reef transcends atmospheric boundaries of land and air. From the frustrations of the inability to film beneath the sea, Hurley created a new figure of environmental allure: the elusive and deceptive power of the ocean's surface, its "skin." Focused on one of Frank Hurley's seaphotographs, Elias gives words to the uncanny power of this threshold zone of tension. At the ocean's "skin," flamboyant corals' partial emergence gave them almost a surrealist existence in photos where the viewer feels the corals return her gaze. Below the watery threshold – which Hurley failed to transcend – lie hazardous, ship-wrecking, and uncanny matters. At the same time, Hurley's attention is distracted, or abstracted, toward analogy: the ocean's reefy roof, in the tropical Torres Strait, recalls Antarctic ice-forms. Undersea aesthetics consistently perform this sort of choreography: inherited structures accrue new senses while having their bounds established with unusual force.

In Sean O'Hanlan's chapter, the maritime notes of surrealism cohere into a frontier of imaginative exploration. The surrealist both created and came to explore the wreck, O'Hanlan writes memorably, embedding the surrealist figures of wreck and salvage diver in the context of a long fantasy lineage of the undersea, but also World War I maritime destruction. O'Hanlan brings out the coordination and tensions between a sense of modernity as shipwreck – and the pursuit of salvage and renewal, which the surrealists took to the depths. Seeking to unite diver and shipwreck, she highlights surrealist leader André Breton's fascination with underwater optics, when Breton in 1925 went to see *The Wonders of the Sea* by J.E. Williamson, inventor of underwater cinema. Again, the chapter integrates the undersea's challenge to visuality as a thread, for Breton found it to unsettle conventional perception in bracing ways and open new imaginative possibilities. Even technologies used to penetrate the sea depths became sources of defamiliarization on land in their surrealist deployment. Thus Salvador Dali's near asphyxiation when he appeared to give a lecture in a diving suit at the International Surrealist Exhibition in London in the summer of 1936.

For the surrealists, physical perception of the distortions of the undersea were weirdly invigorating; however, with the advent of the ability to see wreckage on the ocean floor, undersea exhibition took a tragic turn. Margaret Cohen's article traces new fantasies of wreckage enabled by such technology, in the first picture of shipwreck based on observation of submarine conditions by British maritime painter William Lionel Wyllie, and then in the filming of wreckage on the ocean floor by Jacques-Yves Cousteau and Louis Malle in *The Silent World* (*Le Monde du silence*) (1956). Wreckage connects with the romantic enjoyment of picturesque ruins in Wyllie's *Davy Jones's Locker* (1890), and is transformed into a gothic spectacle, adapting cinematic techniques to organize these haunted spaces that belie the Enlightenment rule of reason. Another thread of Cohen's essay running throughout the volume is the inventive imagination on the part of documentarians, facing the need to shape into comprehensible form an unprecedented, disorienting environment. Like O'Hanlan's surrealists, Wyllie, Cousteau and Malle find disquieting yet undeniable potential in haunted submarine space.

Linda Williams attends to the impact of the Anthropocene on the oceans in her essay on the connections between the black zones of the ocean and our cities of light. Williams draws her examples from NASA photography and from one of the most powerful terrestrial exhibition forms for the undersea: the museum, in an era before photography and film, but also today. She reminds us too of the historicity of ocean consciousness inscribed in its exhibitions, as in the London Natural History Museum's rehanging of a Blue Whale renamed 'Hope,' recently moved to a commanding position at the entrance and animated into a more dynamic diving posture. Williams herself performs a similar sea change, asking us to note the materiality of a statue of Charles Darwin, which, she reminds us, is fittingly wrought from the skeletons of microscopic marine beings, as is so much stone in the city. Like the surrealists in O'Hanlan's account, Williams awakens readers to a *dépaysement* where they feel most firmly located, with the goal not of existential but rather environmental awakening.

Josh Wodak's essay ruminates on marine-ecological and other atmospheres across a five-decade slice of British and American pop music, from the Beatles to Radiohead. Some early manifestations feature the submarine as metaphor for a sort of personal refugia, where it remains possible to engage "childish wonder." Some of these appearances have unanticipated careers: "Strawberry Fields" becomes the name of a site at Guantánamo Bay, "Karma Police" the handle for a GCHQ surveillance program. Do these echoes represent cynical perversions or the movements of signifiers within a distressingly coherent cultural domain? Wodak animates a transforming, Anthropocenic undersea as a creeping presence, traveling among scientific reports and pop culture.

Most remarkable of all may be the volume's uncanny (and unscripted) return, in its final pages, to concerns that animated its early reaches. Janet Laurence's installation artwork *Deep Breathing: Resuscitation for the Reef* is conveyed to readers by photos of its parts and wholes, as well as by Prudence Gibson's accompanying text. Arranged to evoke a spectral hospital ward, *Deep Breathing* assays "an aesthetics of care" for the Great Barrier Reef and the lives, forms, and colors that issue therefrom. It operates, in part, by unsettling: numerous hybrid "obscenities" suggest weird intimacies – and perhaps empathies – among spectators, bodies, cultures, practices, and histories. One key referent is the early modern cabinet of curiosities, where naturalia and artificialia intermingle, sometimes in ways that subsequent eras came to regard as grotesque. But *Deep Breathing* also seems to hearken to Rodríguez Rincón's grottos, which mediate odd encounters among disparate knowledges, aesthetics, beliefs, and forms. This suggests, in a manner alternately edifying and disquieting, that the undersea has always opened spaces for thinking and acting across dissolving boundaries, spaces that have never been ours to master, but may be our obligation to reconceive.

* * *

Notes

1 Adrienne Rich, "Diving into the Wreck," in *Diving Into the Wreck: Poems 1971–1972* (New York, W.W. Norton, 1973), 22. Subsequent citations from the poem are from this edition.
2 Joseph Addison, *The Spectator* 412 (1712): 138.
3 See Robert Doran, *The Theory of the Sublime from Longinus to Kant* (Cambridge, UK: Cambridge University Press, 2015), 180.
4 Edmund Burke, *A Philosophical Enquiry into the Origin of our Ideas of the Sublime and Beautiful*, ed. Adam Phillips (Oxford: Oxford University Press, 1998), 44.
5 Elizabeth Kolbert, *The Sixth Extinction* (New York: Henry Holt and Company, 2014), 139.
6 See Laurence Le Dû-Blayo and Olivier Musard, "Towards a Shared Language: Semantic Exchanges and Cross-disciplinary Interaction," introduction to *Underwater Seascapes: From Geographical to Ecological Perspectives*, eds. Olivier Musard, Laurence Le Dû-Blayo et al. (Cham: Springer, 2014), 1–15.
7 "R.M.S. Titanic, Guidelines," NOAA Office of General Counsel, https://www.gc.noaa.gov/gcil_titanic-guidelines.htm (accessed May 2018).
8 William Abberley, email to Margaret Cohen, June 2, 2018.

Further reading

Note: Pressed for space, we include only one representative title by each critic, and do not cite scholars in the collection.

Adamowsky, Natascha. *The Mysterious Science of the Sea, 1775–1943*. New York: Routledge, 2016.

Alaimo, Stacy. "Jellyfish Science, Jellyfish Aesthetics: Posthuman Reconfigurations of the Sensible." In *Thinking with Water*, eds. Cecilia Chen, Janine MacLeod, and Astrida Neimanis. Kingston, Ontario: McGill-Queen's University Press, 2013.

Anderson, Jon and Peters Kimberley. *Water Worlds: Human Geographies of the Ocean*. Farnham and Burlington: Ashgate, 2014.

Bachelard, Gaston. *Water and Dreams: An Essay on the Imagination of Matter*, trans. Edith R. Farrell. Dallas: Dallas Institute of Humanities and Culture, 1999 (3rd edition).

Clarke, David. *Water and Art*. London: Reaktion Books, 2010.

Clark, Marting and Farquharson, Alex. *Aquatopia; The Imaginary of the Ocean Deep*. London: Tate Publishing, 2013.

Cusack, Tricia. *Framing the Ocean, 1700 to the Present: Envisaging the Sea as Social Space*. Farnham and Burlington: Ashgate, 2014.

Special issue of *Comparative Literature*, "Oceanic Routes: An ACLA Forum," 69.1 (2017). See notably Kerry Bystrom and Isabel Hofmeyr's introduction, "Oceanic Routes: (Post-It) Notes on Hydrocolonialism," 1–6; Alice Te Punga Somerville, "Where Oceans Come From," 25–31; and Elizabeth Deloughrey, "Submarine Futures of the Anthropocene," 32–44.

Dion, Mark. *Oceanomania*. London: Mack, 2011.

Ecott, Tim. *Neutral Buoyancy: Adventures in a Liquid World*. New York: Grove Press, 2002.

Firebrace, William. "Eyes Aquatic." *AA Files* 62 (2011): 47–62.

Helmreich, Stefan et al. *Sounding the Limits of Life*. Princeton, NJ: Princeton University Press, 2015.

Mentz, Steve. *At the Bottom of Shakespeare's Ocean*. New York: Continuum, 2009.

Musard, Olivier, Le Dû-Blayo, Laurence, Francour, Patrice, Beurier, Pierre, Feunteun, Eric and Talassinos, Luc. *Underwater Seascapes: From Geographical to Ecological Perspectives*. Cham: Springer, 2014.

Neimanis, Astrida. *Bodies of Water: Posthuman Feminist Epistemology*. London: Bloomsbury, 2017.

Rozwadowski, Helen *Fathoming the Ocean: The Discovery and Exploration of the Deep Sea*. Cambridge, MA: Belknap Press, Harvard, 2008.

Starosielski, Nicole. "Beyond Fluidity: A Cultural History of Cinema Underwater." In *Ecocinema Theory and Practice*, eds. Stephen Rust, Salma Monani, and Sean Cubitt. New York: Routledge, 2012.

Susik, Abigail. "Convergence Zone: The Aesthetics and Politics of the Ocean in Contemporary Art and Photography." *Drain Magazine*, "Supernature," 7.2 (2012): n.p. http://drainmag.com/convergence-zone-the-aesthetics-and-politics-of-the-ocean-in-contemporary-art-and-photography/

Torma, Franziska. "Snakey Waters, or: How Marine Biology Structured Global Environmental Sciences," in *On Water: Perceptions, Politics, Perils*, eds. Agnes Kneitz and Marc Landry. RCC Perspectives (2012) 2: 13–21.

Von Mallinckrodt, Rebekka. "Exploring Underwater Worlds: Diving in the Late Seventeenth-/Early Eighteenth-Century British Empire." In *Empire of the Senses: Sensory Practices of Colonialism in Early America*, eds. Daniela Hacke and Paul Musselwhite. Leiden and Boston: Brill, 2018. 300–322.

1 The aesthetics of the early modern grotto and the advent of an empirical nature[1]

Luis Rodríguez Rincón

Of the diversity of imaginative aquatic spaces that Gaston Bachelard and Margaret Cohen describe – from fresh to salt water, calm to stormy seas, open oceans versus shorelines – grottos are a little-theorized and historically contingent category.[2] Grottos were a pagan architectural topos that was widely adopted in European architecture and literature starting in the fifteenth century. What complicates any modern assessment of the early modern grotto is the need to suspend divisions that today maintain the study of art separate from the study of nature. In fact, the grotto's flourishing is a testament to the productive energies unleashed by the convergence of pagan aesthetics with natural history in the early modern period. What, then, was a grotto? Why did early modern potentates across Europe invest vast sums to build this pagan architectural topos?[3]

In the sixteenth century, whenever a character was led beneath a body of water they undoubtedly entered a grotto.[4] As representations of what lay below the surface, of what linked oceans to terrestrial sources of water, grottos like the one designed by the Dutch artist Jacques de Gheyn II (1565–1629) were artificial caves showcasing marvels of hydraulic engineering alongside collections of classical artwork and natural specimens. The closest homologue today in terms of meaning and prestige to the grotto would not be the pleasure caves of the rich and famous but rather the modern museum.[5] De Gheyn's grotto (Figure 1.1, see plate section), built in the 1620s for Prince Maurice of Orange-Nassau (1567–1625) at his palace in The Hague, was the Netherlands' first.[6] No doubt inspired by the botanical gardens at the University of Leiden, which de Gheyn drew in 1600, his grotto showcased an expensive array of shells, coral, and sea life and was part of a larger plan for a garden of exotic plants and an aviary.[7]

What baffles modern viewers in this grotto is the co-existence of the empirical with the fantastic in de Gheyn's design. As Claudia Swan argues, de Gheyn's oeuvre straddles the divide between two very different eras of Dutch art: "His is an oeuvre committed equally to the naturalistic representation of the world as it appears and a world of spectral fantasies."[8] In his grotto, naturalistic renderings of real animals and shells appear alongside hybrid creatures that conflate the terrestrial with the oceanic, the organic with the inorganic, and the human with the non-human. The grotto is symmetrical along a vertical axis demarcated by a human figure typically identified as Neptune. He sits encased in billowing shells that merge his humanoid body with the landscape it ostensibly controls. For Felice Stampfle, Neptune's metamorphic dismemberment can be traced directly back to the grotesque style of ornamentation adopted in early modern Europe from rediscovered Roman murals.[9] Neptune's material encasement suggests a sentience to the aesthetics of this underwater realm underscored by the monstrous visage of his throne that emerges like a double-image from between the sea god's legs. The whole landscape seems capable of returning the viewer's gaze. One is rightfully left searching for the line dividing nature from art in a space predicated on the very erasure of such a distinction as the artist demonstrates nature's truth as aesthetics.

Such was the grotto as a submarine microcosm distilled from the interpretation of a pagan topos. While grottos share a distinctive rustic aesthetic that imitated the look and feel of natural forms, it was one that resisted the modern notion of aesthetics. That is, the idea of beauty for its own sake, of surfaces without hermeneutic depths. Rather, the revival of the grotto in the renaissance and its evolution as the sixteenth gave way to the seventeenth century serves as a testament to the productive, though idiosyncratic, early modern linkage of the study of nature to hermeneutics, aesthetics to meaning, and antiquity to the natural world. In sum, the grotto's evolution in the sixteenth century links the early modern advent of an empirical science of nature to a fascination with pagan art.

The conceptual genealogy of the renaissance grotto harkens back to the Greco-Roman practice of the museum in its original etymology as a shrine to the Muses, patron deities of the arts. The ancient Greek practice of designating natural caves and springs as divine spaces was one point of origin for this tradition.[10] Over time such natural, sacred caves came to include artificial components. The construction of artificial grottos that mimicked the features of natural springs spread to Rome by the 1st century BCE, becoming popular with both citizens and emperors alike.[11] With the transition from Roman to medieval times, new grottos stopped being built while old ones fell into ruin or disappeared, leaving a heterogeneous heritage of scattered ruins and textual fragments that to this day reads opaquely. Renaissance grottos appropriated and misinterpreted this pagan tradition by drawing inspiration from a slew of different pagan structures: some religious, some profane, intended for public display, or private enjoyment. Even today, what might be translated as *grotto* conflates different classical structures, whether they be temples, fountains, or baths.[12]

The renaissance revival of the grotto begins in fifteenth-century Italy. The famed Italian polymath Leon Battista Alberti (1404–72) provides the earliest and perhaps most influential early modern description. In his *De re aedificatoria*, published posthumously in 1485, Alberti set the parameters for the design and aesthetic of the grotto for the next century:

> To their grottoes [*antris*] and caves [*criptis*] the ancients used to apply a deliberately roughened revetment of tiny pumice chips, or Travertine foam, which Ovid called 'living pumice.' We have also seen green ocher used to imitate the bearded moss of a grotto. Something we once saw in a grotto gave great delight: where a rushing spring gushed out, the surface

> had been made up of various seashells and oysters, some inverted, others open, charmingly arranged according to their different colors.[13]

What delighted Alberti was the "deliberate" fabrication of an "imitated" nature through the revival of Greco-Roman architectural practices. Delight for Alberti was produced both through the imitation of natural colors and surfaces and through the symmetrical arrangement of shells to produce mosaics that outdid nature's beauty. This was a space of revelation where shells represented the treasures of a submarine world whose empirical reality fascinated curious minds. Alberti's interest in techniques that make nature into a pagan aesthetic practice illustrates a paradox of late medieval and early modern thought: that the interpretation of pagan art animated the study of nature.[14]

The grotto's rustic aesthetic was inseparable from a pedagogical imperative that did not value pagan art and natural beauty for its own sake. Rather, the function of aesthetic delight was to educate an audience by engaging the senses in order to elevate the mind. Alberti's description of a grotto is situated in the section of his treatise where he codified the architectural parameters of the Renaissance villa upon a pagan model.[15] His architectural precepts went hand in hand with his appraisal of a lifestyle based on *otium*.[16] For renaissance intellectuals like Alberti, civilized life was bifurcated between *negotium*, the business characteristic of city life, and *otium*, which celebrated a leisurely rural life dedicated to intellectual pursuits and moral edification. Immediately preceding Alberti's description of the grotto is a list of other delightful ornaments appropriate for the country villa:

> We are particularly delighted when we see paintings of pleasant landscapes or harbors, scenes of fishing, hunting, bathing, or country sports, and flowery and leafy views. It is worth mentioning the emperor Octavian, who collected rare and enormous bones of huge animals as an ornament to his house.[17]

Key to the aristocratic ideal of *otium* was a mixing of delight with an investment of time and money into intellectual pursuits like building, collecting, and studying.[18] Take for example the Marchesa of Mantua Isabella d'Este's (1474–1539) grotto, built by 1508, where "adjacent to Isabella's studiolo, the small anticamara of the grotto served as a museum providing a combination of *studium* and *otium* ... adorned with allegorical paintings."[19] Intended as a space of respite and reflection, the grotto was from its Italianate beginnings associated with collecting classical objects and natural curiosities for study and admiration: "with its massing of curiosities and precious stones and in its cramped quality, the grotto overlapped with the later *Wunderkammer*."[20] This holds true for the first grottos built around the Papal Court, which displayed the growing collection of Greco-Roman statuary in a rustic garden setting. Donato Bramante (1444–1514), the architect who designed the Belvedere Courtyard at the Vatican, with its fountain-niches built to display the recently rediscovered Roman statues of the Nile and Tiber, was also responsible for designing a nymphaeum built in 1508–11 for Cardinal Pompeo Colonna in Genazzano.[21]

The pedagogical imperative implicit in renaissance notions of delight and *otium* linked the aesthetic of the grotto to a hermeneutic practice that saw meaning in their alluring beauty. Jean Seznec schematizes the way in which pagan mythology and art was transported and transformed from antiquity to the renaissance by Christian readers and artists. The hermeneutic rehabilitation of pagan mythology in the hands of the early church fathers divorced the classical forms of Zeus, Hercules, and Neptune from their interpretation. In sum,

> form and subject survived in isolation, so to speak, each distinct from the other. As pagan ideas gradually became severed from expression in art, Christian ideas came forward to inhabit the forms thus abandoned, just as Christian cult took over the empty temples or the imperial baths.[22]

For Seznec, what made the renaissance distinct from the Middle Ages was the reintegration of classical forms with their original signification: "the Renaissance appears as the reintegration of antique subject matter within the antique form."[23] What complicates matters is that pagan mythology had accrued in the hands of ancient writers as distinct as Plato and Cicero.[24] What the renaissance did then to the grotto was to revive it as a hermeneutic vehicle, a *fabula*, for conveying a hidden truth that was open to competing interpretations. While Alberti described the aesthetic form of the grotto, the meaning connected to this form had a different genealogy than its aesthetics.

The meaning attributed to the grotto as a *fabula* in the early modern period owes a great deal to the Neoplatonist allegory codified by Porphyry (233–305 CE) in his commentary on Homer's *Odyssey*. A disciple and editor of his contemporary Plotinus (204–70 CE), Porphyry's impact on early modern Neoplatonists like Pico della Mirandola (1463–94) and Marcilio Ficino (1433–99) is substantial. Porphyry's *On the Cave of the Nymphs*, which is arguably the oldest extant piece of literary criticism, reconciles Homer's imagery with a Neoplatonist cosmology.[25] Homer describes a grotto as Odysseus finally reaches Ithaca, disembarking from the Phaeacian vessel into a sacred sea-cave described as follows:

> and at the head of the harbor is a slender-leafed olive
> and near by a lovely and murky cave
> sacred to the nymphs called Naiads.
> Within are kraters and amphoras
> of stone, where bees lay up stores of honey.
> Inside, too, are massive stone looms and there the nymphs
> weave sea-purple cloth, a wonder to see.
> The water flows unceasingly. The cave has two gates,
> the one from the north, a path for men to descend,
> while the other, toward the south, is divine. Men do not
> enter by this one, but it is rather a path for immortals.
> (13.102–12)[26]

For Porphyry, Homer's grotto was a microcosm representing the drama of souls as they enter and leave the material world. The cave is described as both "lovely and murky," expressing the duality of existence as a synthesis of idealized, i.e. "lovely," forms and consuming, i.e. "murky," matter:

> Since matter itself is in a state of flux and in itself is deprived of that form through which it takes shape and is made manifest, they took the dampness and humidity of caves, their darkness and, as the poet says, 'murkiness,' as an appropriate symbol of the properties the cosmos owes to matter.[27]

The ontological dichotomy of the lovely and the murky, spirit and water, form and matter, structures Porphyry's interpretation of the grotto: from the bees drawn to honey (human souls drawn to the sensuality of matter), to the nymphs weaving sea-purple cloth ("the sea-purple cloth would clearly be the flesh, woven of blood . . . the wool is even dyed with a product derived from living creatures").[28] Porphyry's allegorical reading of Homer's grotto was arguably better read than the *Odyssey* itself in Europe prior to the sixteenth century.[29] His interpretation resonated strongly with renaissance thinkers convinced the artistry of pagan poetry could be reconciled with Christianity to better understand the mysteries of God's creation.

With Porphyry's reading as a model, the grotto's renaissance revival was energized by a hermeneutic tradition prevalent in the late medieval study of natural history which interpreted pagan *fabula* as conveying truths about the natural world. Leonard Barkan notes in the medieval assessment of pagan stories a tendency to interpret them as *fabula* illustrating *physica ratio*, natural processes, thus grounding an understanding of natural phenomena in the study of pagan art.[30] The discipline of natural history, starting in the twelfth century, was animated by poetic hermeneutics as the methods of interpretation of the latter defined the truths of the former. Medieval naturalists, in sum, "used pagan myths as essential data in the mapping of the cosmos."[31] Myths and nature existed as reflections of each other:

> Since so many of the myths were from their classical beginning grounded in descriptions of nature, the corpus of myths was for medieval humanists a ready-made poetization of the cosmos that they wished to scrutinize. So *fabula* is generally inseparable from *physica ratio*, not only because the fables are grounded in etiologies and interpreted as nature allegories but also because they exist in the newly rediscovered pagan world of the Goddess Natura.[32]

The special hermeneutic status of pagan *fabula* made aesthetics central to the illustration of natural truths. The form of the grotto and its humanoid denizens of nymphs and tritons were interpreted in the renaissance as divinely inspired veils behind which lay an understanding of natural truths.

The hermeneutic meaning attributed to grottos as *fabula* in the sixteenth century is evidenced by Natale Conti's (1520–82) *Mythologiae* (1567), an authoritative reference for knowledge about pagan mythology until the seventeenth century.[33] Conti's compendium professed to interpret pagan *fabula* "to bring to light the philosophical principles shrouded in the stories: I mean those principles that reveal the powers and actions of nature, shape morals, establish an honest way of life, or disclose the motions and powers of the planets."[34] When Conti turned his hermeneutic gaze to the interpretation of grottos and their humanoid inhabitants, nymphs, what he found were secrets of natural history:

> The nymphs were supposed to live underneath the ground, because the ancients thought that the source of sweet water was hidden away in subterranean caverns, and was composed of air that had been changed into water, just as sea vapors were changed into rain. And since moisture's power, which helps things to reproduce, was filtered through the sea, rivers, lakes, and fountains, rivulets, and mountains, the ancients made the Nymphs the guardians of all those places.[35]

Conti's willingness to interpret natural truths behind the veil of grottos emphasizes the vitality of the hermeneutic knot conflating *fabula* with *physica ratio* throughout the sixteenth century. The grotto as an ornate abode of nymphs was for Conti a way of representing the supra-human order that governed the elemental world.

The grotto as *fabula* dominates Francesco de' Vieiri's (1524–91) Neoplatonist interpretation of the grottos at Pratolino. Designed by the Florentine polymath Bernardo Bountalenti (1531–1608) for the Grand Duke of Tuscany, Francesco de' Medici (1541–87), Pratolino, completed by the early 1580s, was an apex of the Italian Mannerist style of landscaping.[36] Pratolino's grottos deeply impressed contemporary visitors like Michel de Montaigne (1533–92), who considered it "surpassing all that we had seen elsewhere" during his Italian sojourns of 1580–81.[37] Montaigne's written account describes the multitude of sensorial pleasures experienced by visitors: "Not only does the movement of water produce music and harmony but also hydraulic forces power automated statues, doors, mechanical animals that move to drink water and similar machines."[38] Vieiri's Neoplatonist interpretation links Montaigne's description of Pratolino's sensorial delights to transcendent cosmographic truths.

Vieri's commentary, *Della Maravigliose Opere di Pratolino, & d'Amore* (1586), elucidates the meaning hidden behind Pratolino's material pleasures. As a *fabula*, "a fiction, where human or divine wisdom has hidden some useful truth under the disguise of the marvelous or delightful," Vieiri interpreted Pratolino like Porphyry did the Homeric grotto.[39] Its purpose was to delight guests while communicating the esoteric nature of the material world. Each grotto and vista helped to elucidate such truths:

> From the stairs one sees a path forty arms in length, around which here and there are many fountains shooting forth jets of water forming a pergola over one's head that does not wet the passer-by. By this long and wide path, sloping downwards, is represented the whole course of human existence.[40]

Looking down a path lined with fountains, Vieiri saw represented mortal life with its physical and moral dangers. The sight, sound, and potential touch of water communicated the vicissitudes of material existence to an enlightened visitor:

> This path we judge to be terribly long due to the many varied troubles that accompany it, and because such a journey separates us from the promised peace of beatitude. Those jets of water are the predictable miseries of this world, which we labor to avoid, the which are deemed lowly and thus we keep the middle path of virtue to avoid such dangers.[41]

For Neoplatonists like Vieiri and Porphyry, the grotto was the material world made symbolically legible as a hermeneutically charged aesthetic. Water's sensorial delights represented the earthly entanglements to be avoided by those seeking spiritual transcendence. The goal was to transcend the grotto of sensual delights for the spiritual world, "the promised peace of beatitude." Though Vieiri's Neoplatonist reading of the grotto was influential, his did not exclude other interpretations.[42]

The hydraulic marvels that delighted Montaigne – the water organs, moving statues, and shooting jets of water – while of symbolic importance to a reader like Vieiri, represented an empirical understanding of water's physical nature that – as the sixteenth gave way to the seventeenth century – had diminishing use for the hermeneutics of Neoplatonism. Phillip Morel highlights a general shift in grotto design over the course of the sixteenth century from an antiquarian focus to a proto-scientific interest in depicting natural processes:

> from the first grottos, marked by a humanist and antiquarian spirit, we pass on to solutions in which there begins to appear a 'naturalist' preoccupation, which subsequently becomes the characteristic and predominant trait of Florentine grottos of the last third century . . . now it manages to reflect nature in its becoming and to compete with it in its effects. This attitude, typically Mannerist, leads, in the case of the grotto, to a truly analytic approach, an investigation into the internal mechanisms of nature, and an exegesis of its hidden laws.[43]

Grottos marveled as windows into the secrets of nature, but a nature increasingly understood on the basis of methodologies distinct from those of poetic hermeneutics. The desire to build grottos, to re-create a pagan architectural topos, fueled an interest in investigating the empirical reality of the natural world in order to better represent it. The changing epistemological framework guiding grotto design in the sixteenth century highlights how nature began to speak its own, empirical language, but one still in the service of aesthetics. For more empirically-oriented builders, like the Frenchman Bernard Palissy (1510–89), the goal was no longer to create a *fabula* for nature that symbolically conveyed truth but to re-create nature itself.

Palissy is remembered primarily for his rustic ceramics, but he holds a special place in the history of the philosophy of science for his explicit rejection of classical authorities in favor of his own practical experience as an artist and a natural historian.[44] In the Preface to his *Discours admirables* (1580), he seeks:

> to silence those who ask, "how it is possible for a man to know something and speak of natural effects without having seen the Latin books of the philosophers?" Such a discussion could happen in my studio since through practical demonstrations I prove wrong many theories of philosophers, even the most famous and ancient, as anyone can see and hear in less than two hours, if they are willing to take the trouble to come see my cabinet, in which one can behold marvelous things placed there as proof and testament of my writings.[45]

A desire to study nature itself as recreated in his cabinet characterizes Palissy's approach to representing nature.[46] His empiricism was grounded in his faith as a Huguenot who believed in the unadulterated study of the Bible.[47] His understanding of nature was as much a product of his artistic labors as his famed ceramics; his rustic aesthetic was nature's empirical truths made malleable to human hands and legible to reverent eyes.

In Horst Bredekamp's account of the conjoined early modern history of aesthetics and natural history, the emergence of an empirical interest in understanding nature was part and parcel of the recuperation of pagan art: "The idea that works of art, in particular the art of antiquity, could mediate between human beings and nature was a basic tenet of both the natural sciences and aesthetics."[48] With its links to the better-known *Wunderkammer*, the grotto was a theater in which these conjoined but seemingly contradictory impulses were gathered in productive ways. For Bredekamp, it is perfectly reasonable that a grotto-builder like Palissy would merge an antiquarian aesthetic to an empirical naturalism since the grotto "was a perfect location in which to manifest the transition of apparently untouched yet structured nature to art in the style of antiquity and finally to automatons brought to life, since the grottos were viewed as anthropomorphic 'wombs' where metals became more highly developed, as though in an underground 'laboratory.'"[49] For Palissy, the meaning of the grotto does not go beyond its physical realization. The value of its aesthetic is precisely in standing-in for the reality it reproduces.

In his written description of a grotto built for Anne de Montmorency (1493–1567), Palissy claims to create not a representation of nature but the thing itself in his grotto:

> While I contemplated these figures and ideas, the inventor of the grotto said to me, "Does this work seem beautiful to you?" "Yes," I said, "and very strange." He replied, "It is stranger than you know. Because the lines, figures, and shapes that appear on the surface are also visible on the inside." I expressed my disbelief at what he claimed. Then he took a piece

> of one of his porcelain statues and broke it into pieces, so I could see the truth of his words. Everywhere one could see lines and forms in the broken pieces of ceramic, just as if it had been a true rock. I found that very strange and amazing.[50]

The relation between the surface and interior of a work of art is suspended by Palissy's claim to produce artefacts that are themselves seemingly a product of the natural world. This negates the grotto's status as a *fabula* since its meanings no longer need go beyond its sensorial nature. Nature's truth is palpable in the grotto as nature is ostensibly reproduced by the artist. Another French Huguenot, Salomon de Caus (1576–1626), follows in the footsteps of Palissy in his treatise on grottos.

The illustrated title page of de Caus's treatise (Figure 1.2), *The Causes of Moving Forces with Diverse Machines both Useful and Pleasing to Which are Added Many Designs for Grottos and Fountains* (1615), highlights the classical roots of the mechanical arts and their utility in a modern context.[51] In the foreground are depicted the two great engineering minds of antiquity, Archimedes (c.287–c.212BCE) and Hero of Alexandria (10–70CE), surrounded by levers, pulleys, and other instruments of the mechanical arts. Immediately above them are the gods Mercury and Vulcan, patrons of crafts such as medicine and ironworking. Between the gods is a slate upon which is written the title of the work tilted open at an angle to reveal a nautical scene in the background. It is at sea where the utility of Salomon's exposition of physical forces is demonstrated as ships sail across oceans while cranes load cargo onto ships, all harnessing the same physical forces as the grottos he describes. That this nautical scene exists outside of the classical architectural facade of the title, beyond the pagan gods and their great engineering minds, indicates the author's desire to surpass the ancients and their knowledge.

In his dedication to Louis XIII of France (1601–43), de Caus argues there is a need for rulers to surround themselves with educated advisors, "but also that they themselves familiarize themselves with such knowledge, especially of mathematics and other dependent sciences, so as to not fall prey to the false council of falterers."[52] Mathematics is thus a new necessity for the modern-day ruler. In his Introduction, Salomon goes further by making the revival of the pagan arts, by which he means the mechanical arts, a central achievement of the great rulers of the previous century. The allegorical rehabilitation of classical mythology characteristic of an initial phase in the renaissance revival of antiquity – its revalorization as *fabula* conveying the occult wisdom of antiquity – is erased in de Caus's retelling of the renaissance adoption of the pagan arts:

> And some time after the city of Rome was destroyed by the Goths, who destroyed the most beautiful works of Italy, and after the people of Europe paid little attention to the arts, until the reign of three great princes, which all lived during the same time, the Emperor Charles V, the French King

Figure 1.2 Salomon de Caus, title page, *Les Raison des Forces Mouvantes*(Frankfurt: Jan Norton, 1615). Courtesy of the Smithsonian Libraries, Washington, D.C.

François 1er, and the King of England Henry VIII, all of which demonstrated great curiosity in reestablishing the arts so long hidden.[53]

De Caus's use of the term 'art' refers primarily to the mechanical arts, but his treatise connects antiquity's knowledge of hydraulics and mathematics to the aesthetic practices of the grotto. His discussion of grottos indicates the bond linking aesthetics to nature was beginning to fray by the seventeenth century.

In de Caus's design for a grotto depicting Polyphemus and Galatea (Figure 1.3), the machinery takes on a prominence that eclipses the pagan forms they animate. He describes a "Machine in which is represented a Galatea, driven by two dolphins in a loop, all the while above her a Cyclopes plays a reed flute."[54] In de Caus's description, it is the technical detailing of gears and water pressure rather than the mythological figures which delight and instruct his readers. In fact, there is no discussion of Galatea and Polyphemus as *fabula*. The grotto's rustic aesthetic was itself such a convention that de Caus only briefly mentions, as an aside, that "when it comes to the ornament of said grotto, it can be made from various rocks, and rustic shells, or with frames of figures or grotesque decorations."[55] In the image that accompanies this description, the machinery dominates the representation. The pagan *fabula* of Galatea and Polyphemus, which once fascinated Vieiri and Porphyry, are now puppets tied to the marvel of gears, pulleys, and hydraulics.[56]

Figure 1.3 1.1 Salomon de Caus, design for a grotto depicting Polyphemus and Galatea, in *Les Raisons des forces mouvantes* (Frankfurt: Jan Norton, 1615). Courtesy of the Smithsonian Libraries, Washington, D.C.

De Caus claims the gods Vulcan and Mercury were once human inventors of machines like himself who were later deified by those who mistook their ingenuity for magic.[57] If the pagan gods once demonstrated their power through the creation of machines, then moderns like himself can claim an equal if not greater prestige for themselves through their mastery of natural forces. When seen in conjunction with de Gheyn's grotto from the Netherlands, what is revealed is how seemingly contradictory impulses from a modern epistemological perspective – like empiricism versus a fascination with pagan artwork as nature – productively intertwined in the early modern period. If the aesthetic of the grotto came to be valued by the seventeenth century more for the empirical information its fabrication conveyed than for its symbolic meaning, that does not negate the fact that the desire to build grottos, to re-create nature as seen through the filter of antiquity, once catalyzed an empirical investigation of nature. Like de Gheyn's grotto discussed earlier – which was built around a decade after de Caus published his treatise – these historically divergent epistemological impulses co-existed productively in the early modern grotto before their Enlightenment divorce.

De Gheyn's morose and enshelled Neptune recalls the Neoplatonist reading of the grotto as a microcosm of the material world in need of spiritual transcendence. The decomposing bodies that surround the central figure remind viewers of the dichotomy between their transient material form and their eternal spiritual nature. This symbolic message co-exists with de Gheyn's empirical interest in depicting aquatic life in a naturalist vein while also harnessing hydraulic forces to pour water out the two amphoras on either side of Neptune and all along the bottom ridge. The early modern grotto represents a time when the artistic and philosophical fascination with a pagan architectural topos fueled a nascent empirical interest in the mysteries of the submarine and their relation to ultimate questions of human meaning and purpose. If many artistic and scientific trends that follow after the early modern grotto gesture towards a unity between aesthetics and science, it is because the drive to represent the submarine carries nascently the desire to understand the alluring shapes, textures, and colors of our terraqueous world.

Notes

1 All translations are the author's unless otherwise indicated.

2 Gaston Bachelard, *L'eau et les rêves: essai sur l'imagination de la matière* (Paris: J. Corti, 1942). Margaret Cohen, "The Chronotopes of the Sea," in *The Novel*, ed. Franco Moretti, vol. 2 (Princeton: Princeton UP, 2006), 647–67.

3 Italy alone counts hundreds of early modern grottos catalogued in the two volumes of the *Atlante delle grotte e dei ninfei in Italia*, eds. Vincenzo Cazzato, Marcello Fagiolo, and Maria Adriana Giusti (Milan: Electra, 2001). For a discussion of grottos in England and France see Roy Strong, *The Renaissance Garden in England* (London: Thames and Hudson, 1979) and Naomi Miller, *Heavenly Caves: Reflections on the Garden Grotto* (New York: G. Braziller, 1982).

4 Examples of the grotto as a prototypical underwater space abound in early modern literature, from Jacopo Sannazaro's *Arcadia* (1504) to Torquato Tasso's *Gerusalemme liberata* (1581) in Italy. In Castilian, grottos appear in Garcilaso de la Vega's poetry

(1543) and fascinate the later Cervantes as well. In Portugal and England, both Luis de Camões and Edmund Spenser describe grottos when tasked with imagining underwater spaces in their respective epics.

5 Paula Findlen, "The Museum: Its Classical Etymology and Renaissance Genealogy," *Journal of the History of Collections* 1, no. 1 (1989): 60. Horst Bredekamp, *The Lure of Antiquity and the Cult of the Machine: The Kunstkammer and the Evolution of Nature, Art and Technology*, trans. Allison Brown (Princeton: Markus Wiener Publishers, 1995), 49–51. Martin Kemp, "Wrought by No Artist's Hand: The Natural, the Artificial, the Exotic, and the Scientific in Some Artefacts from the Renaissance," in *Reframing the Renaissance: Visual Culture in Europe and Latin America 1450–1650*, ed. Claire Farago (New Haven: Yale University Press, 1995), 196.

6 I.Q. Van Regteren Altena, *Jacques de Gheyn: Three Generations: Text*, vol. 1 (The Hague: Martinus Nijhoff Publishers, 1983), 141–2.

7 Like tulips soon after, shells in early modern Europe were a hot commodity considered signs of wealth and prestige. I.Q. Van Regteren Altena, *Jacques de Gheyn: Three Generations: Catalogue*, vol. 2 (The Hague: Martinus Nijhoff Publishers, 1983), 46. Claudia Swan, *Art, Science, and Witchcraft in Early Modern Holland: Jacques de Gheyn II (1565–1629)* (Cambridge: Cambridge University Press, 2005), 58–9.

8 Swan, 22–3.

9 Felice Stampfle, "A Design for a Garden Grotto by Jacques de Gheyn II," *Master Drawings* 3, no. 4 (Winter, 1965): 381–3.

10 G.W. Elderkin, "The Natural and the Artificial Grotto," *The Journal of the American School of Classical Studies at Athens* 10, no. 2 (1941): 125. Jennifer Larson, *Greek Nymphs: Myth, Cult, Lore* (Oxford: Oxford University Press, 2001), 226. The famous Oracle at Delphi is the best-known example.

11 Elena Cenci, "'L'antro delle ninfe': significato ed evoluzione dall'Antichità al Cinquecento," in *Il Giardino Storico All'Italiana*, ed. Francesco Nuvolari (Milan: Electra, 1992), 147. Hervé Brunon and Monique Mosser, *L'Imaginaire des grottes dans les jardins européens* (Paris: Hazan, 2014), 16, 20. Miller, 18. The massive grotto complex at Sperlonga, north of Naples, testifies to the prestige of such buildings for Roman Emperors as does Hadrian's Villa at Tivoli.

12 Georges Roux, "Le Val des Muses, et les Muses chez les auteurs anciens," *Bulletin de correspondance hellénique* 78 (1954): 45. Cenci, 149. Miller, 17.

13 Leon Battista Alberti, *On the Art of Building in Ten Books*, trans. J. Rykwert, N. Leach, and R. Tavernor (Cambridge, MA: MIT Press, 1988), 299.

14 Leonard Barkan, *The Gods Made Flesh: Metamorphosis and the Pursuit of Paganism* (New Haven: Yale University Press, 1986), 117, 120–2. Bredekamp, 13.

15 Brunon and Mosser, 291–2.

16 Strong, 14–15. David R. Coffin, *The Villa in the Life of Renaissance Rome* (Princeton: Princeton University Press, 1979), 9–14.

17 Alberti, 299.

18 A.D. Fraser Jenkins, "Cosimo de' Medici's Patronage of Architecture and the Theory of Magnificence," *Journal of the Warburg and Courtauld Institutes* 33 (1970): 162–70.

19 Miller, 44.

20 Ibid., 44.

21 Miller, 43. Strong, 17.

22 Jean Seznec, *The Survival of the Pagan Gods: The Mythological Tradition and Its Place in Renaissance Humanism and Art*, trans. Barbara F. Sessions (New York: Pantheon Books, 1953), 212–13.

23 Ibid., 211

24 Jason Crawford, *Allegory and Enchantment: An Early Modern Poetics* (Oxford: Oxford University Press, 2017), 62.

25 Robert Lamberton, *On The Cave of the Nymphs* (Barrytown, NY: Station Hill Press, 1983), 3.
26 Ibid., 24.
27 Ibid., 24.
28 Ibid., 29–37.
29 Ibid., 13.
30 Barkan, 123.
31 Ibid., 121.
32 Ibid.,121.
33 Seznec, 248–51.
34 Natale Conti, *Mythologiae*, trans. John Mulryan and Steven Brown, vol. 1 (Tempe, AZ: Arizona Center for Medieval and Renaissance Studies, 2006), 2.
35 Ibid., 391.
36 Miller, 49. Strong, 78.
37 Michel de Montaigne, *Journal de voyage*, ed. Fausta Garavini (Paris: Gallimard, 1983), 175.
38 Ibid., 175.
39 Francesco de' Vieiri, *Discorsi di M. Francesco de' Vieri, detto il Verino secondo, cittadino fiorentino, delle marauigliose opere di Pratolino, & d'amore* (Florence: Appresso Giorgio Marescotti, 1586), 67.
40 Ibid., 45.
41 Ibid., 45.
42 Harry Berger, Jr., *Second World and Green World* (Berkeley: University of California Press, 1988), 36.
43 Philippe Morel, "Mannerist Grottos in Sixteenth-Century Italy," in *Sixteenth-Century Italian Art*, ed. Michael W. Cole (Malden, MA: Blackwell, 2006), 121.
44 Kemp, 191–3. Bredekamp, 49.
45 Bernard Palissy, "Discours Admirables," in *Oeuvres complètes*, ed. Keith Cameron, Jean Céard, Marie-Madeleine Fragonard, Marie-Dominique Legrand, Frank Lestringant, and Gilbert Schrenk (Paris: Honoré Champion, 2010), 250.
46 Leonard N. Amico, *Bernard Palissy: In Search of Earthly Paradise* (Paris: Flammarion, 1996), 7.
47 Ernst Kris, *Le Style rustique*, trans. Christophe Jouanlanne (Paris: Macula, 2005), 133–4.
48 Bredekamp, 12.
49 Ibid., 49.
50 Bernard Palissy, "Ordonnace de la grotte rustique," in *Oeuvres complètes*, ed. Cameron, Céard, Grafonard, Legrand, Lestringant, and Schrenk (Paris: Honoré Champion, 2010), 78–80.
51 Salomon de Caus, *Les Raisons des forces mouvants avec divers Machines tant utilles que plaisants aus quelles sont adioinct plusieurs desseings de grottoes et fontaines* (Frankfurt: Jan Norton, 1615), Title Page.
52 Ibid., Au Roi.
53 Ibid., Epistre.
54 Ibid., livre 1, prob. XXIII.
55 Ibid., livre 1, prob. XXVI.
56 Bredekamp, 50–1.
57 *Caus, Les Raisons des forces mouvants*, Epistre.

2 The porcellaneous ocean

Matter and meaning in the rococo undersea

Killian Quigley

Off the coast of Connacht, in the west of Ireland, lie large beds of a coralline algal (*Lithothamnion*) residue called maerl. It is one of the ocean's many apparent, or actual, oxymorons. Formed, like pearls, from calcium carbonate, and bearing close resemblance to coral, maerl derives, in fact, from a marine plant. It can provoke a sense like the one Ed Ricketts and John Steinbeck recalled feeling in the Gulf of California, when they encountered plumularian hydroids. Those "animals" were, "in appearance at least . . . so like plants" that they seemed to "indicate to the imagination a bridge between flora and fauna."[1] For the writer and cartographer Tim Robinson, deposits of Irish maerl suggest other, and perhaps odder, bridges. He describes them as "composed of tiny twiglike bits of something like unglazed pottery, white, cream coloured, pale green or faintly violet flushed."[2] They remind him, in other words, of branching plants, and of earthenware. Met with uncommon combinations of shape, texture, and color – with uncommon lives – Robinson's language can only approximate, and its imprecision is as obvious as it is poetic. Thus a small but signal instance of a tenacious phenomenon: marine stuff has long challenged – and long thrilled – writers, artists, and scientists who search terrestrial epistemologies and memories for points of reference, coordinates that often fail to accommodate stony plants and ceramic animals.

In Robinson's pottery-talk, this chapter recognises kinship with a long-standing, strange, and seminal literary and visual motif. Underwater shapes and lives are frequently metaphorized in terms of decoration, and especially via reference to the plastic arts. In *The Log from the Sea of Cortez* (1951), Ricketts and Steinbeck described beautiful, and exceedingly agile, crabs called Sally Lightfoots (*Grapsus grapsus*), whose "brilliant cloisonné carapaces" skittered about the foreshore.[3] "Cloisonné," a term for a technique in enamel-making, whereby the parts of an image are separated into compartments – as if jointed, we might say – and separately colored. This is overtly aesthetic, but it would be over-simple to say that it's unscientific. Ricketts was a path-breaking authority on littoral invertebrates, and co-author of *Between Pacific Tides* (1939), a key text for students of the Californian intertidal. And science hasn't stopped thinking in these terms since. In pursuit of an "ocean ethic," the marine ecologist Peter J. Auster and colleagues urge the revelation of the ocean's "biological treasures" to a potentially, but not yet actually, engaged public.[4]

If it's true that making the undersea more perceptually available might in turn galvanise greater concern and better care for it, then the decorative topos is worth trying to understand. To do so, it will be useful to engage with an exuberant period in that topos's life, a period which coincided with significant changes in Western natural philosophy's understanding of the undersea.[5] In eighteenth-century Europe, the submarine was made visible via dazzling objects and ornaments, rendered in porcelain, silver, jewels, stucco, and other materials besides. The art and literature that expressed this stuff, and this sensibility, is often called rococo, a rather loose term for a playful, sensual, and frequently ostentatious mode that was delineated with particular vigor by the plastic arts. Accused, in the eighteenth century and on sundry occasions since, of insalubrious ostentation, femininity, and superficiality, rococo forms have not infrequently provoked widespread – and remarkably efficacious – critical condemnation. Paradoxically, but vitally, a significant strain of subaqueous meaning is nonetheless detectable in the vibrations among rococo sense, early marine science, and the matter and lives that informed them.

The "arrival in art of the entire sea bed" – thus has the art critic Waldemar Januszczak described the subject of his pastel-hued pilgrimage, *Rococo: Travel, Pleasure, Madness* (2014).[6] Januszczak's formulation is apt and arresting, but he doesn't really ponder what the Rococo's sea-floor sensibility means *for* the undersea, or for understandings thereof. In the late seventeenth and eighteenth centuries, ocean lives and ocean spaces posed imaginative problems for poetic and aesthetic conventions like the pastoral, the picturesque, and the sublime. Those problems did not confine themselves to literature and art: they bore upon all attempts to ideate the sea. This was true, not least of all, of the proto-oceanographic researches of such natural philosophers as Robert Boyle, Luigi Ferdinando Marsigli, and Jean-André Peyssonnel. As Januszczak suggests, the Rococo manifested not just a pronounced enthusiasm for shell-curved lines and coralline forms, but a sustained marinal sensibility, one eminently suited to domains that could be – and often still are – described as repositories of extraordinary treasure.

The rococo ocean's ostentation testified to the productive difficulties posed by things, and approximations of things, that did not quite conform to emergent Enlightenment taxonomies, or to conventional accounts of relations among animate and inanimate matter. This essay does not posit the Rococo as the resolution to eighteenth-century awkwardnesses in marinary knowledge, let alone the resolution to our own. Nor, for that matter, does it contend that decorative sea-sense surged, ex nihilo, from the rococo eighteenth century. Rather, it argues that the Rococo furbished a submarine sensibility that drew deeply from past modes, but was aesthetically and epistemologically productive – in an era that histories commonly describe as Enlightened by the natural sciences[7] – in ways that have rarely been acknowledged. And above all, this chapter recognises and attends to the active principle in rococo submersion, that dynamic indeterminacy that flourishes, like porcelain coral, where senses meet the undersea.

The rococo view

Toward the end of an essay called "The Temperature of the Subterranean and Submarine Regions, As to Heat and Cold" (1671), the seventeenth-century Anglo-Irish natural philosopher Robert Boyle explained that water at the sea floor is remarkably tranquil. He framed this claim in terms of aesthetic disruption, or inaccessibility: "This calmness of the sea will appear strange," he writes, "to many, who, admiring at the force of stormy winds, and the vastness of the waves they raise, do not, at the same time, consider the almost incomparably greater quantity and weight of water that must be moved, to make any great commotion at the bottom of the sea."[8] As for the sea-ground itself, and what sort of topography it might present, Boyle consults diverse accounts, but perceives little of consensus. In certain spots, such as Mannar, in what is now Sri Lanka's Northern Province, the bottom of the ocean seems intriguingly varied, but in the majority of cases this is true only near shore; at depth, the lie of the sea floor seems to change only gradually.[9]

In the first case, Boyle's account seems to preemptively defuse the potential for sea-bottom sublimity, or at least derange its character: several decades later, Edmund Burke would argue that "the great and sublime in *nature*" produces "astonishment," that "state of the soul, in which all its motions are suspended, with some degree of horror."[10] Burke also claimed "obscurity" for one of the sublime's central features,[11] and the deep ocean strikes us now, as then, as offering obscurity in the extreme, but its apparent tranquility makes it something less than – or at least different from – sublimely terrible in the terrestrial sense. The picturesque, another prominent eighteenth-century technique for visioning Nature, might fare even worse: the irregularities, undulations, agitations, and variety that William Gilpin idealized in picturesque landscapes – and even in the surface of the ocean[12] – seem frustratingly absent from Boyle's calm, and mostly flat, abyssal plain.

Other aesthetic structures were on offer. During the first two-thirds of the eighteenth-century, there flourished a tendency – in the plastic arts, architecture, painting, music, and literature – that later eras would come to know, and often to disparage, as the Rococo. As spirit and practice, it is no more receptive to hasty summary than any other movement, and its bounds remain the subject of definitional contestation today. Vividly recognizable, however, are its enthusiasms for form, surface, delicacy, asymmetry, and extremities of recherché artfulness and artisanship. Insistently oriented toward materiality and sensuousness, the Rococo is perhaps most accurately – if incompletely – understood as a proclivity for ornament, or as an energy that becomes visible when ornamentation diverts significant attention from the things adorned.

"Rococo" comes from the French *rocaille*, which refers originally to the pseudo-natural arrangement of shells, rocks and stones in and around garden grottos. Famous for shelly, florid, scrolling forms that seem to melt or flow seamlessly from one to the next, the Rococo is often described in terms of liquidity. Take, for instance, François Boucher's curious screen design *Rocaille*, the primary effect of which we might understand as a wave (Figure 2.1).

The engraving is filled with some things we might expect to be cast up as jetsam by the tide, as well as many things we might not. Generally, compositionally speaking, we can perhaps imagine Boucher's picture as a single wave, swelling leftward and upward across the picture. The odd and frankly nonsensical distortions visible in the temple in the background might even tempt us to wonder whether this is an undersea image. In any case, we see here – if we have the patience to distinguish them – corals, sea-sponges, and some examples of what is perhaps the Rococo's central obsession, the shell.

By no means absent from sculpture and painting, the Rococo's most superb creations are nonetheless decorative. Its curvaceous transformations of material direct the eye like so many instances of the English artist William Hogarth's line of beauty.[13] Calling for more freedom of ornamentation in architecture, Hogarth pointed to "shells and flowers" as a stock "of elegant hints for this purpose."[14] Still more striking is Hogarth's elaboration of a generalized theory of shelly looking. He suggests that objects of aesthetic contemplation be imagined as emptied of their contents, and reduced to exterior shells – this, to enable a perspectival fantasy in which the eye can examine form from within, as well as from without. This insists on a robust spatialization of aesthetic regard, not only insofar as the thing admired is imagined to have an inside and an outside, but as the observer becomes acutely aware of the physical relation between their body, their eyes, and an object.[15]

The Rococo frustrates interpretation by resisting narrativization, preferring to seduce its viewer into lingering forever over and around the surfaces of things. It shares, in this sense, a sensitivity for description, for detail, and for surface with seventeenth-century Dutch painting, in contrast to the iconographical tendencies of the Italian Renaissance.[16] Looking at three-dimensional decorative forms invites the eye to ecstatically expand its possible orientations, and informs it – often brusquely – that it will not manage to see in all ways at once. Even when the Rococo does not dispense altogether with narrative information, it at least redefines what "reading" a work of visual art might involve. Exemplary of this tendency are the cartouches, splendid objects of attention in their own right, which, at times, brazenly contain nothing whatsoever (Figure 2.2). Some rococo cartouches radically dispense altogether with an inside, compelling their viewers not only to abandon conventional modes of looking, but to indulge the fullness of their delight in the mobile, and perhaps interminable, circuits of decorative shape.

As neoclassicism strengthened its claims to the high ground of taste, the Rococo and its near relations were tidied into ontological niches like fashion, decoration, gardening, and interior design, which connoted triviality and, as Barbara Maria Stafford notes, the feminine.[17] The Rococo was also closely associated with music, spectacle, *trompe l'œil*, and the theater, making it a ready target for critics

Figure 2.1 Claude-Augustin Duflos after François Boucher, *Rocaille,* 1730s, etching, engraving on white paper, 49.4 x 24.8 cm. Cooper Hewitt, Smithsonian Design Museum.

Figure 2.2 Anonymous French artist, *Rococo Cartouche,* 18th century, black crayon, 27.9 x 19.4 cm. The Metropolitan Museum of Art.

who regretted the eighteenth century's appetite for sensation.[18] One of numerous recent critics to recuperate the Rococo, Jennifer Milam reorients its "ludicity" to signal fertile ground for creative experimentation: eighteenth-century play could expand aesthetic possibilities through immersive participation and sensory dynamism. Milam argues that Immanuel Kant's celebration of caprice and sensation in French art may indicate his debts to the Rococo – his ideal of "purposiveness without purpose" may have drawn inspiration from the epicenters of bodily and sensory ludicity, eighteenth-century pleasure gardens and rococo decoration.[19] Kant's valorization of the individuated, or bounded, natural aesthetic object does seem to find its corollary in rococo décor, and in the domestic spaces adorned thereby. But if the Rococo's *objets trouvés* might have satisfied Kant's eye, they clearly failed to signal purposiveness without purpose in the eyes of their detractors, who maligned their lack of order or intention.[20]

Porcelain at the sea-bottom

To fully grasp the Rococo's oceanic visions, it is worth dwelling upon the style's relationship to its greatest forebear, the Baroque. The latter word derives from the Portuguese *barocco*, which designates an irregularly-shaped pearl.[21] The Baroque is literally bejeweled by animal forms, some veracious – lions, rabbits, eagles, ostriches, and cranes – and others fabulous. While the Baroque sometimes looked directly to nature – notably via Flemish printmakers – it often did so in search of emblems of strangeness, and of other worlds. Baroque artists who worked with pearl, such as Erasmus Hornik, took formal direction from the ocean: their works elaborated upon whatever shapes an unfinished specimen suggested. Nature's shapes spurred not verisimilitude but fantasy, and baroque pearl-work results from a shimmering interplay between design, artisanal skill, and found form.[22]

The Rococo inherited from the Baroque this paradoxical impetus, to at once honor and exploit raw nature and to explore the marvelous, monstrous, and capricious possibilities it discovered. But if the Baroque is the visual countertype of a million-organed choir, the Rococo is akin to the pianoforte. Baroque profusion is thick and weighty; rococo abundance is built from fish-bones. And *Rococote* bizarrerie is a consequence not so much of the natural forms employed, but of those forms' superabundance, and their uncanny juxtaposition with apparently incongruent entities. While hardly monolithic – the Rococo does sometimes involve itself in didacticism and moralizing – its jumbled nature as frequently challenges any search for system or regulation.[23]

Dolphins, sea monsters, and pearls indicate the Baroque's marine dreams, but the Rococo is still more enthusiastically immersed. A 1741 terracotta medallion and stand by Jean-Baptiste Janelle the younger exemplifies this tendency (Figure 2.3, see plate section). Cast "directly from nature," the coralliform support attracts by far the greater share of the viewer's attention; the portrait medallion itself (and this is one of a pair) is, by contrast, generic. Janelle, who died in 1764, is an almost total mystery – besides this piece's mate, which is housed with it at the Rijksmuseum in Amsterdam, only one other work, a watch-stand which lives at the Louvre, has been confidently attributed to him.[24] Janelle's stand prompts us to wonder whether we can describe the Rococo's undersea aesthetic as verging, in some cases, on overwhelming traditional forms and genres, or at least upsetting expected symbolic hierarchies. Instead of presenting an identifiable person for honor, glorification, or satire, Janelle commemorates coral, shells, and an elegant lobster.

Through an ewer and basin from the Capodimonte factory in Naples (c. 1750), porcelain enables the influx of the undersea into domestic display (Figure 2.4, see plate section). In addition to the shells and other encrustations that cover them, both pieces have been elaborately rendered in order to appear as though they could have been found at the bottom of the sea. Perhaps the viewer is being invited to imagine them as uncannily familiar natural phenomena, requisitioned for human use. Or maybe they're more like the hybrid results of an ewer and a basin which, having fallen to the sea floor in the course of a shipwreck, became something other and more than themselves.

The status of the sea-bottom – and the things that sank there – was pondered not only by rococo designers, but by contemporary legal theorists. Lagan, explained John Exton's *Maritime Dicæologie* (1746), designates that matter "which lyeth on the Sea-ground, or is taken from the bottom of the Sea."[25] If

Janelle's coralliform stand showed the potential for the rococo undersea to overwhelm standard form and genre, the porcelain ewer and basin indicate an even more striking – and, depending on one's sensibilities, perhaps quite garish – example of this possibility. We observe, again, the strange and arresting simultaneity of scrupulous attention to natural detail and a kind of ecstatic creativity. It isn't altogether difficult to understand why some contemporaries described this sort of thing as mad.

Materials matter, and among the Rococo's numerous media, porcelain's place is especially important. In a perspicacious analysis of the collections of the Duchess of Portland, and of the auction thereof in 1786, Stacey Sloboda interprets an odd commingling of natural-historical and porcelainic specimens. In a manner that reminds, strikingly, of Ricketts and Steinbeck's hydroids, and of Robinson's maerl, Sloboda explains that "porcelain played a particular semiotic role within the collection itself, providing a conceptual bridge between the natural and the artificial, and thus acting as a unifying element among its aesthetically and conceptually disparate parts." Many of the natural objects on offer were shells, and some of the ceramic pieces, such as a "carp vase," evoked underwater lives. The auction's various stuffs, and the manner in which they were juxtaposed, amounted to thematic connection between "ceramics and the marine world." Those connections were international, too: as Sloboda points out, porcelain presences also signified the importance of Chinese and Japanese traditions in making.[26]

The Duchess of Portland's collections comprised one, but by no means the only, instance of rococo natural history. In France, Boucher's *Rocaille* was used as the frontispiece for a landmark work of eighteenth-century conchological science, Dezallier d'Argenville's *Histoire naturelle éclaircie*. E. C. Spary demonstrates that, in an era that had not universally claimed to disentangle connoisseurship from scientific inquiry, shells and the books that pictured them were at least as appealing to aesthetes as they were to natural historians. Or more pointedly: the aesthetic and scientific work of volumes like these were inextricable from one another. Taking this view, there is no contradiction in apprehending the conchological literature of Antoine-Joseph Dézallier d'Argenville's conchology as productive, simultaneously, of rococo art and scientific knowledge. D'Argenville's contemporaries were not unanimously untroubled by this, but the decorative presentation of even such ostensibly anti-aesthetic texts as Michel Adanson's *Histoire naturelle du Sénégal* (1757) confirms the Rococo's unremitting presence.[27]

The Rococo offers a modicum of nuance to narratives of an agonistic eighteenth century in which classification spars with connoisseurship, and curiosity with science. Then and since, the Rococo has frequently been described as having a complex and often troubled relationship to the Enlightenment, which Max Horkheimer and Theodor Adorno influentially described as "the disenchantment of the world." In Enlightenment, Horkheimer and Adorno observe an order manufacturing validations of itself, like a species following a reproductive imperative to establish and defend a biological niche. Perhaps, by these lights, the ocean was unamenable to the program because it was not perceptible in terms of "mere objectivity."[28] But rococo worlds are other than eccentrically unnatural. The nineteenth-century German architect Gottfried Semper described the Rococo through organicist metaphor: boiserie, he wrote, elevated the frame to the status of an acting "*organism*," which enveloped what it contained as "*a plant*" might, potentially disordering its arrangement.[29] Refracted by Semper's attitude, a second look at Janelle's design, and at the Capodimonte vessels, may yield a glimpse of enchanted organisms.

Ornamental behavior

Among the most significant, and paradigm-shifting, developments in eighteenth-century oceanic natural philosophy were the contributions of the naturalist and *marseillais* Jean-André Peyssonnel. Revising the pioneering work of Luigi Ferdinando Marsigli, and of Boyle, Peyssonnel had successfully disproven the dominant definition of coral – that it was a sort of flowering plant – by identifying the "insects of the vermicular kind" – "*urtica, purpura,* or *polype*" – that construct it.[30] The English physician William Watson promoted Peyssonnel's work to the Royal Society in London, where a summary of the Frenchman's findings were published at midcentury. Though previously unprinted, Peyssonnel's work had by then achieved sufficient continental currency that the entomologist René-Antoine Ferchault de Réaumur wrote to him in 1726 to express reservations about his findings.

Boyle had clearly imagined the undersea through reference to terrestrial vegetation, identifying marine plants as related, if oddly, to their counterparts on land. He had it, for instance, from a "man of learning" in the Maldives that "a sort of cocoa-trees" grow on the sea floor there, the fruit of which is "real," though smaller "than most other sorts of cocoa's [*sic*]." Boyle's informant speculates that the underwater tree has sprouted from immersed nuts of the land-lubbing variety; this is a theory of undersea flora as subaqueous deviation from a terrestrial norm. Submerged, the fruit is easily cut and eaten, but after exposure to the air, it hardens extremely.[31] This sort of idea recalls Ovid's tenacious vision of coral as supple underwater and rigid in air, a misconception that wasn't corrected in Europe until the Sicilian botanist Paolo Boccone pushed out in a fisherman's boat and, reaching under the surface, encountered hardness.[32]

Boccone's discovery did not entail that coral is incapable of metamorphosis. As Peyssonnel knew, it does transform, regularly and dazzlingly, when its "insects" protrude from their orifices in search of food.[33] And Peyssonnel explains, indeed fixates upon, coral's fascinating potential for strange architectonic behavior. He had discovered, in 1721, that because corals do not behave like plants, they flourish upon, and actually incorporate, diverse materials. Where Boyle suggested underwater forests and gardens, Peyssonnel presented subaqueous *rocaille*. Coral grows, he reports, in ways that confuse the boundary between natural and artificial, between the built and the ground. His vermicular insects are seen to behave "like shell-animals," and inhabit a "stony" sea-ground.[34] For Peyssonnel, coral "roots" are "ligneous" and "stonish" (*pierreuses*), and can grow – thanks to insect generation – on any solid body whatever.[35] So within coral branches, one finds such bodies as rocks, shells, and even bits of broken bottles and pots.[36]

Marion Endt-Jones has defined coral "as a networked object," one "that epitomizes the connectivity not only of nature, but also of different fields of creativity and knowledge production – of *ars* and *techné*, art and craftsmanship, art and nature, and nature and representation."[37] This chapter bears her thesis out, and enhances it, too. Endt-Jones argues that the Enlightenment and its inheritors have always operated through "philosophical and epistemological binaries," and that corals have always resisted them.[38] Its hybridity, she claims, made it uniquely suited to pre-Enlightenment practices in aesthetics and knowledge, such as cabinets of curiosity. Endt-Jones is, of course, right – but the connectivities she describes, and the weird commensurability of coralliferous art and decorative animals, suggest that even supposedly Enlightened ocean-going science may be underwritten by an uncooperative attitude toward dualisms.

After all, with Peyssonnel's findings in view, *coquillier* encrustations on buildings, stoups, and medallion-stands appear something other than ridiculous. This is not to say that Peyssonnel consciously interpreted his findings through the frame of decorative style, or that *rococoterie* had hold of his mind, or his eye. It is, however, to recognize in retrospect the significant congeniality obtaining between early eighteenth-century empirical views of coral behavior and rococo matter. It is tempting to recognize Peyssonnel's sea as an array of decorated surfaces, full up with the sort of "garnish" that Robert Morris, author of the Palladian *Select Architecture* (1755), decried in the Rococo.

> [Our] modern Architects . . . have made Ornament or Dress, the principle Part of their Performance, and have given Decoration to ill-proportion'd Fabricks, and indeed, Superfluity is generally the thing to attract the eye; they garnish the inelegant Design, to attone [sic] for the Disproportion of the Parts, and crowd and fill the Spaces by some gay Dress, to conceal the want of Proportion; which is only a kind of unmeaning Attempt at Elegance.[39]

Superfluity denotes not only extravagance and excess, but excrescence, and exuberant growth. Morris's polemic prompts consideration of whether more-than-human architects, such as corals, have been held, wittingly or otherwise, to strictures like these. As for concealment, the undersea is unusually nutritive of deception. In Peyssonnel's account, the sea's productions had – and the location of agency is significant here – tricked some into thinking they were all stones and minerals, and others into believing them all to be vegetation.[40] From these errors, Peyssonnel explains, terminological mistakes ensued, leaving all tree-like undersea life lumped together as coral and the rest scattered in the ledger as madrepores, "lytophitons," and "alcionions."[41]

Later in the century, William Cowper's "Retirement" (1782) indicated the extent to which art could draw upon sharply bifurcated senses of the submarine, and endorse one while castigating the other. The poet's stated aim is "To dive into the secret deeps within,/To share no passion and no favourite sin,/And search the themes, important above all,/Ourselves, and our recovery from our fall." For Cowper – as for Percy Shelley, several decades on[42] – the undersea is useful as a symbol of a plumbless self, a rangeless space for examination and illumination. But "Retirement" also reprimands superciliousness and venality through the figure of the rococo aesthete, who treads the "dangerous shore" for "shining pebbles" here, and "weeds and shells there." Like Morris, Cowper's speaker implies that glitter has the power to seduce out of reason and into harm's way: "The waves o'ertake them in their serious play,/And every hour sweeps multitudes away."[43] The point is not to establish that Cowper was some reactionary, because he wasn't. It is, rather, to illustrate the multifariousness of the sea's aesthetic meanings, and the multiple ways those meanings have fared. And it bears saying that, reading "Retirement," it's possible to imagine peering closer at the errants and recognising Ricketts, Steinbeck, and Robinson, remarking and marvelling at an intertidal tableau.

Rococo growths

If Glenn Adamson is right, and "rococo art and design implied that reality itself was manipulable through the techniques of artifice," then it is incumbent upon students of eighteenth-century culture to hail the involvement of oceanine practice. Peyssonnel's vision of coral growth finds an outstanding, far-flung correlate in rococo phenomena at the American Porcelain Factory, which opened in 1770 in Philadelphia. Shells were employed to form clay molds, which were used in turn to cast *coquilles* in porcelain. These then attached themselves to other objects, in whatever organization and quantity suited.[44] The *porcelainier*'s studio began to operate like a porcellaneous ocean, as the identity of the art object migrated toward lifelikeness. Jean-Baptiste Janelle's medallion and coralliform stand are the products of a similar method. Duplicative behavior was widespread among rococo things: Janelle's terracotta was likely to create copies of itself in faience, bronze, or porcelain.[45]

Rococo visions of ingenious oceanic form and dazzling undersea treasure remain available today, from the aesthetics of the fish market to the BBC's *The Blue Planet* (2001) and *Blue Planet II* (2017). This prompts us to complicate David Clarke's idea that "the modern attitude toward water" issued, to an overwhelming extent, from "certain technologies in nineteenth-century art, especially of the Romantic period."[46] *Sub*aqueous color, sound, time, space, and scale operate differently than their terrestrial analogues. Humans have difficulty establishing shape, and relative size, because of a loss of contrast and a lack of peripheral visual stimulation. Things tend to look larger and closer than they are in fact.[47] Bringing the undersea into view tends to require exceptionally aggressive imaginative and technological intervention, as standard visuality is of little use in a realm that is overwhelmingly lightless.[48] Signs become signals of human sensory lack, or of human sensory neglect. When photography and diving made the undersea accessible in novel ways, many discoveries looked like confirmations of rococo imagination: coral "grottoes" supported "a never-ending parade of beauty."[49] And to a vital extent, coral reefs *are* all surface: beneath brilliant, burgeoning extremities, their substrates are skeletons.[50]

To suggest a positive connection between underwater worlds and an ostensibly antirational representative system like the Rococo is not to imply that the undersea is or ought to be beyond science's ken. But it is to ponder the forms of detection which engage those parts of the globe that natural philosophy has struggled, at specific junctures, to apprehend. It is also to imply that when such forms are underestimated or shut down, the space they leave behind is not always filled by alternative, superior schemas. An analysis like this one runs the risk of essentializing the eighteenth-century ocean as alien and unknowable, as a fundamentally fantastic site. But from a scholarly milieu that has sometimes accepted that the eighteenth century was at best terrified of the ocean or at worst unaware of its existence, it is worth taking other visions seriously. Perhaps "nature" has never been adequate to the ocean.

In a simple but significant sense, we might say that one of the valuable – and potentially actionable – provocations recognizable in these sources is that the undersea has proven productive and problematic even for those cultures that spent little time getting into it. More particularly, we might apprehend, in our own space and time as well as in others, the flourishing of language, image, and culture at the borders where human ingenuity meets material contingency and earthly incitement. Many of the stories we tell about European Enlightenment retrospectively attach themselves, limpet-like, to texts, styles, and works of art that appear to welcome dualistic accounts of the relationship between what we choose to call science and whatever we separate therefrom. This chapter contributes to the voices of those who challenge the biform view. The eighteenth-century Rococo offers a rich and underappreciated insight into some more intertwisted trajectory in the era's thinking. We ought to recognize it as a species of sea-tangle; if this can be seen to be true, then underwater realms emerge, at times murkily, at times spectacularly, as diversely nutritive for a singular moment in the histories and futures of feeling, knowing, and form.

Notes

1 John Steinbeck, *The Log from the* Sea of Cortez (London: Penguin, 2000), 194.
2 Tim Robinson, *Connemara: The Last Pool of Darkness* (Dublin: Penguin, 2008), 283.
3 Steinbeck, *The Log from the* Sea of Cortez, 53.
4 Peter J. Auster et al., "Developing an Ocean Ethic: Science, Utility, Aesthetics, Self-Interest, and Different Ways of Knowing," *Conservation Biology* 23, no. 1 (2009): 233–5.
5 Rebekka von Mallinckrodt is one critic recently to explain that the era in question staged witnessed increasing "empirical experience" of the submarine. See Von Mallinckrodt, "Exploring Underwater Worlds: Diving in the Late Seventeenth-/ Early Eighteenth-Century British Empire," in *Empire of the Senses: Sensory Practices of Colonialism in Early America*, eds. Daniela Hacke and Paul Musselwhite (Leiden and Boston: Brill, 2018), 301.

6 *Rococo: Travel, Pleasure, Madness*, directed by Waldemar Januszczak (London: British Broadcasting Corporation, 2014), DVD.
7 See, for example, Rémy G. Saisselin, *The Enlightenment Against the Baroque: Economics and Aesthetics in the Eighteenth Century* (Berkeley: University of California Press, 1992), 5–6.
8 Robert Boyle, "The Temperature of the Subterranean and Submarine Regions, As to Heat and Cold," in *The Philosophical Works*, ed. Peter Shaw, vol. 3 (London: Innys et al., 1725), 246.
9 Ibid., 243–4.
10 Edmund Burke, *A Philosophical Enquiry into the Origin of our Ideas of the Sublime and Beautiful*, ed. Adam Phillips (Oxford: Oxford UP, 1998), 53.
11 Ibid., 54, 58.
12 Gilpin offers a lapidary excursus on the picturesque potential of the sea: "Nothing gives so just an idea of the beautiful swellings of ground, as those of water; where it has sufficient room to undulate, and expand. In ground, which is composed of very refractory materials, you are presented often with harsh lines, angular insertions, and disagreeable abruptnesses. In water, whether in gentle, or in agitated motion, all is easy; all is softened into itself; and the hills and the vallies play into each other in a variety of the most beautiful forms. In agitated water abruptnesses indeed there are; but yet they are such abruptnesses, as, in some part or other, unite properly with the surface around them; and are, on the whole, perfectly harmonious. Now if the ocean, in any of these swelling, and agitations, should be arrested, and fixed, it would produce that pleasing variety, which we admire in ground." Gilpin, *Observations on the River Wye, and Several Parts of South Wales, &c.* (London: R. Blamire, 1782), 62–3.
13 See William Hogarth, *The Analysis of Beauty* (London: J. Reeves, 1753), 25.
14 Ibid., 46.
15 Ibid., 7–8.
16 Martin Jay, "Scopic Regimes of Modernity," in *Vision and Visuality*, ed. Hal Foster (Seattle: Bay Press, 1988), 12–13.
17 Barbara Maria Stafford, "The Eighteenth-Century: Towards an Interdisciplinary Model," *The Art Bulletin* 70, no. 1 (1988): 16.
18 Ibid., 18.
19 Jennifer Milam, "Play Between Disciplines: The Problem of the Ludic in Rococo Art and Enlightenment Culture," in *The Interdisciplinary century: tensions and convergences in eighteenth-century art, history and literature*, eds. Julia V. Douthwaite and Mary Vidal (Oxford: Voltaire Foundation, 2005), 104–13.
20 Roger Scruton, *Beauty: A Very Short Introduction* (Oxford: Oxford University Press, 2011), 50–1, 66.
21 Jay, "Scopic Regimes of Modernity," 16; Elizabeth Rodini, "Baroque Pearls," *Art Institute of Chicago Museum Studies* 25, no. 2 (2000): 68. This etymology has been disputed, but the consensus, pearling view is overwhelming. René Wellek argued unconvincingly that the term describes an illogical syllogism. See Wellek, "The Concept of Baroque in Literary Scholarship," *The Journal of Aesthetics and Art Criticism* 5, no. 2 (1946): 77.
22 Rodini, "Baroque Pearls," 68–9.
23 Glenn Adamson, "The Real in the Rococo," in *Rethinking the Baroque*, ed. Helen Hills (Surrey and Burlington: Ashgate, 2011), 145.
24 Reinier J. Baarsen, *Paris 1650–1900: Decorative Arts in the Rijksmuseum* (New Haven: Yale University Press, 2013), 136–7.
25 John Exton, *The maritime dicæologie; or, sea-jurisdiction of England* (London: 1746), 52.
26 Stacey Sloboda, "Displaying Materials: Porcelain and Natural History in the Duchess of Portland's Museum," *Eighteenth-Century Studies* 43, no. 4 (2010): 455–72.

27 E. C. Spary, "Rococo readings of the book of nature," in *Books and the Sciences in History*, eds. Marina Frasca-Spada and Nick Jardine (Cambridge: Cambridge University Press, 2000), 262–70.
28 Max Horkheimer and Theodor W. Adorno, *Dialectic of Enlightenment: Philosophical Fragments*, ed. Gunzelin Schmid Noerr, trans. Edmund Jephcott (Stanford: Stanford University Press, 2002), 1–27.
29 Gottfried Semper, *Style in the Technical and Tectonic Arts, or, Practical Aesthetics* (Los Angeles: Getty, 2004), 710.
30 William Watson, "An Account of a manuscript treatise, presented to the Royal Society, intituled, *Traité du corail*," *Philosophical Transactions* 47 (1751–2): 448–54.
31 Boyle, "The Temperature of the Subterranean and Submarine Regions," 249.
32 Watson, "An Account of a manuscript treatise," 449.
33 Jean-André Peyssonnel, *Traduction d'un Article des Transactions sur le Corail* (London: 1756), 26.
34 Watson, "An Account of a manuscript treatise," 454-60.
35 Jean-André Peyssonnel, "Traité du Corail . . . Pour server a l'histoire naturelle de la mer," manuscript, British Library, Add MS 4219, ff. 30–1.
36 Ibid., ff. 49–50.
37 Marion Endt-Jones, "'Something Rich and Strange': Coral in Contemporary Art," in *Framing the Ocean, 1700 to the Present: Envisaging the Sea as Social Space*, ed. Tricia Cusack (Farnham and Burlington: Ashgate), 230.
38 Ibid., 225.
39 Quoted in Adamson, "The Real in the Rococo," 153.
40 Peyssonnel, *Traduction d'un Article des Transactions sur le Corail*, 49.
41 Ibid., 50.
42 In the "Ode to Liberty" (1820), Rome's incipient tyranny is tempered by art's bathic soundings: a "multitudinous anarchy did sweep, / And burst around their walls, like idle foam, / Whilst from the human spirit's deepest deep / Strange melody with love and awe struck dumb / Dissonant arms." Percy Bysshe Shelley, "Ode to Liberty," in *Prometheus Unbound: A Lyrical Drama in Four Acts, with Other Poems* (London: C & J Ollier, 1820), 214.
43 William Cowper, "Retirement," in *The Poems* (New York: Charles Wells, 1835), 147.
44 Adamson, "The Real in the Rococo," 150.
45 Baarsen, *Paris 1650–1900*, 136–7.
46 David Clarke, *Water and Art* (London: Reaktion, 2010), 19.
47 S. M. Luria and Jo Ann S. Kinney, "Underwater Vision," *Science* 167, no. 3924 (1970): 1455–8.
48 William Firebrace, "Eyes Aquatic," *AA Files* 62 (2011): 58.
49 Francis P. Shepard, *The Earth Beneath the Sea* (Baltimore: The Johns Hopkins University Press, 1959), 185.
50 Ibid., 185.

3 The logic of the invisible

Perceiving the submarine world in French Enlightenment geography

Hanna Roman

In the "Preliminary Discourse" (1751) of the French Enlightenment *Encyclopedia, or Reasoned Dictionary of Sciences, Arts and Crafts*, the mathematician Jean D'Alembert described the organization of disciplines as "a kind of map of the world." Each article was like a single country, presumed to be linked geographically and culturally to others, but these links could be difficult to perceive. What obscured them was likened to the ocean: "The Universe is nothing but a vast Ocean, upon the surface of which we perceive ... islands, some bigger, some smaller, whose link to the continent is hidden from us."[1] To create the map of disciplines, seemingly arbitrary islands of knowledge had to be placed within a larger network, but the ocean blurred the pathways that connected them.

This comparison of the order and extent of knowledge to islands in an ocean is, in fact, also one way to approach Enlightenment understandings of the ocean and the submarine realm. D'Alembert depicted its waters as a source of mystery that obscured and prevented the search for truth. Contrary to the commonly-accepted idea that the Enlightenment sciences rejected philosophical, hypothetical systems in favor of empirical observations, in the case of the ocean direct vision and sensory experience would not suffice. The ocean could not be described in and of itself. It was an obstacle to be overcome in the search, through rational and philosophical methods, for what it covered and hid: the fundamental mechanics of the natural world concealed by water. That which was invisible to the human eye was considered more true than initial observations, and the mystery of the ocean had to be removed to access this truth. The ocean thus was not really a thing, but the embodiment of a process of discovery and of a belief that the entirety of nature eventually could be possessed in the terms of human knowledge.

The invisible could not be immediately obtained: to access the underlying truths of nature, the ocean had to be traversed and confronted. The quest for what lay below the ocean's waters occurred through rational method, a step-by-step process of comparison which moved from the visible to invisible, from surface to depth. Writing about the discovery of knowledge in general, D'Alembert charted this movement, describing a method of revelation which proceeded from sensory observations, the beginning of all knowledge, to mental abstractions, which he understood as true knowledge. He sought to "contain in one, unitary system, the infinitely varied branches of human knowledge," and in so doing, to "go all the way back to the origin & the generation of our ideas."[2] From the real extent and dimensions of objects in the world, D'Alembert wrote of refining them to their "figurative area." Knowledge originated in local, historical moments of observation, but it only became ordered, scientific, true, when perception was abstracted. Truth was produced "through the rational study of phenomena, . . . through the art of reducing, as much as will be possible, a great quantity of phenomena to a single one that can be seen as the principle."[3] Like the waters of the ocean, the arbitrary, material form of the sensory world had to be removed, through steps of logical reasoning, in order to access its deeper nature and meaning. True meaning was held, D'Alembert thought, in one single, occult point that united nature and human thought: "the Universe, for the person who would know how to embrace it in a single point of view, would be . . . *but one unique fact & one great truth*."[4]

In presenting the different disciplines that he hoped the *Encyclopedia* would define and explore, D'Alembert not only wanted to show their visible relationships to one another in a tree of knowledge, but also the invisible, rational methods through which they shaped, and were shaped by, human understanding. The Enlightenment study of geography illustrated this process, first discovering the laws of the visible world, of dry land, and then applying these laws to unknown areas such as the undersea. To understand the ocean, or rather what it covered from sight, one had to begin with what could be seen, with what already had a language: islands, cliffs, beaches, shores. Instead of imagining new paradigms of observation, natural philosophers sought to unite visible and invisible through the terms of one language, making isolated islands logically continuous with one another through what they thought of as a universal grammar of nature. The aesthetic of the invisible in the case of geography was one of rational continuity. Through language thinkers gave form to rational analogies, and sought to make real – that is, to realize – the unseen in terms of the seen. As other essays in this volume also show, discourses of the underwater realm often were born at the frontier between known and unknown, established and uncharted, real and conjectural. In the process of speaking the relationships between former and latter, new worlds and modes of perception were invented.

Visualizing the world: finding a language of nature

What did the language of discovering the invisible look like? How were the sea and its mysteries addressed and named? Enlightenment studies of the Earth often spoke of the "terraqueous globe," a planet imagined from above in terms of the broadest and most general relationships between land and water. This philosophy is captured in the article "Physical Geography" in the *Encyclopedia*. Its author, Nicolas Desmarest, described the Earth as seen from a single, distant fixed point; from this perspective the world was resolved into a great system and machine of perpetual motion. Desmarest combined "geography," the simple description of the surrounding terrain, with "physics," or the search for causes and laws, resulting in:

> the rational description of the grand phenomena of the earth, & the consideration of general results deduced from local and particular observations, which have been methodically combined and united into different classes, and inserted into a schema capable of showing the natural economy of the globe, insomuch as it is envisioned only as an uninhabited and sterile mass.[5]

The physical changes the Earth appeared to undergo and the oppositions it seemed to embody – continents vs. oceans, islands vs. lakes, gulfs vs. peninsulas, straits vs. isthmuses – resolved into a single relational point embodying the "natural economy" of the globe, or the continuous equation between all its parts such that each part would recall the meaning of the whole entity. The physical geography of the earth simultaneously portrayed the world in both a broad and general blink of the eye and in the continuous series of all its minute relationships. The world was less a reality than a rational ideal. The Earth, Desmarest wrote, which at first and to untrained eyes "showed us nothing other than the image of a pile of debris and a world in ruins . . . now offers to our enlightened eyes an appearance of order and uniformity, where general relationships unveil themselves in the wake of our footsteps."[6]

Desmarest wrote in part in response to an area of study known as physico-theology, most famously embodied in René Descartes' *The World, or Treatise on Light* (1664) and in the works of English thinkers such as Thomas Burnet (*A Sacred Theory of the Earth*, 1684) and William Whiston (*A New Theory of the Earth*, 1696). These authors had understood the ocean – which they often called the "abyss" or the "subterranean waters" – as a force of change and destruction. The result of the Biblical Flood, which they accepted as a real geological event in the Earth's past, the waters of the ocean were not only symbolic of God's punishment of humankind, but were also blamed for having shattered earthly paradise, leaving this ancient world in ruins. This former world remained only in chaotic fragments: it could be still observed in the irregular shapes of mountains, valleys, and the seabed, from whose chasms and abysses Burnet thought the waters of the flood had erupted. (For Whiston, the waters had originated in the tail of a comet passing over the Earth.)[7] In both cases, the waters were thought to be only temporary, lasting as long as the age of humankind. They hid a meaning that would be revealed upon their retreat, at the end of secular time and the beginning of a new age – a moment often marked by conflagration instead of inundation. The ocean became an interpretative barrier and challenge. To see underneath its waters was to make sense of the past and to predict the future.

Burnet and Whiston, – as well as Isaac Newton (who greatly influenced their thought), – sought to align their theories of the geographical history of the Earth with the events of the Bible. They believed that these two corresponding stories were legible according to the same system of interpretation. They worked to discover the general and universally applicable laws of both nature and religion in the age of human sin, using these laws to explain worlds hidden by time – and, in this case, water.[8] Enlightenment thinkers such as Desmarest and D'Alembert – although they sought to distinguish rational from religious discourse, and to separate the story of the Earth from that of the Bible – appear to have borrowed from the physico-theologians this principle of interpreting the invisible through present-day, visible evidence, speaking of making the natural world rational and of transforming and reducing ruins and mysteries into logical, legible components of a system. In his study of physical geography, Desmarest translated the elements of the Earth into a consistent, uniform, predictable language, in which all elements could be related to one another once the correct interpretative key and perspective had been found. This key would convert the physical world into a series of geometrical relationships gathered together into a single truth. Knowledge of the world became truer as it was rendered more abstract, less visible but consequently clearer and more easily decipherable in its fundamental workings. In a short text on earthquakes, Desmarest wrote that "As single & isolated facts only indicate vagary, I allowed myself to interpret them by bringing them closer together. Their dialogue, their mutual relationships, their union brought forth the idea of mechanism."[9]

A mechanism was a kind of machine, and it was in this term that language, knowledge, and aesthetics merged in the depiction of uncharted worlds such as the submarine. The word "machine" was used in the visual arts at the time, especially to explain about how to view landscape paintings. According to the contemporary definition of "machine" in the *Encyclopedia*, a painting could be described as a machine when, at a distance, all of its elements came together in a "grand organization, grand composition." This composition revealed the "intelligence," the ordering and rational mind, which had created it.[10] Painting and the surveying and comprehension of the geography of landscapes were linked. Andrew Graciano describes the dialogue between representation of landscapes, geology, and mining (another important 'underworld' space) in late eighteenth-century England: the depiction of surface-level rocks "was influenced by the contemporary understanding that the Earth's surface disclosed a great deal about its depths. . . . [T]opography becomes the physiognomy of the land, in which the external appearance is the result of a hidden cause."[11] Because it allowed for the perception and definition of the whole, a painting or rational "mechanism" was often seen as more correct and reliable than the observation of reality itself. It was in this spirit that authors of geography and natural history tended to write about the ocean and of what was concealed below the waves.

Depicting the invisible: rational visions of the undersea

To show how the ocean and what lay beneath it were described, what was the geographical language of the land, of the visible world? Desmarest explained how the world's mountain chains – even islands, which he regarded as the summits of underwater mountain chains – were linked and corresponded to one another. He described mountains from above as a tree, with roots, trunk, and branches:

> The surface of our globe is traversed by several chains of mountains, which link & unite in each continent, & which have a very pronounced correspondence from one continent to the next. These chains embrace, as much through their main trunks as through their parallel branches, almost the entire area of the known surface of the globe. Those mountains which form the principal stems, are the most immense in both their height and mass. They ordinarily occupy & traverse the center of continents.[12]

Natural historians studying the ocean in the early and mid-eighteenth century employed similar language to render the undersea realm visible, logical, and knowable. Many of their texts posited ideas similar to Desmarest's: that all aspects of the physical, terraqueous globe were consistent with one another, because they were parts of the same system. Often geographical phenomena were described as the proportional branches of a tree, or as parts of a human body, with arteries, veins, organs, bones, and muscles that mutually defined one another's purpose to the larger whole: the economy of the entire globe.

In the very controversial but influential *Telliamed, or conversations between an Indian philosopher and a French missionary on the diminution of the sea* (1748), the Enlightenment naturalist Benoît de Maillet devised an account of a fictitious Indian philosopher's (named Telliamed) method for studying the Earth's history. Analogous to dissecting a body, this method uncovered related layers of history and meaning:

> By breaking down the [components that comprise] the substance of this Globe through an exact anatomy of all of its parts, he [Telliamed] first learned of what materials it was composed, & what order these same materials shared. This knowledge, combined with the spirit of comparison always essential to whomever tries to pierce [through] the veils in which nature likes to hide itself, served as a guide for our Philosopher.[13]

De Maillet argued that the current geographical, geological, and biological state of the planet was the result of the activity of water. He did not believe in a single, catastrophic flood like the one in the Bible, but rather spoke of the building up and receding of water over many centuries of time. (These dangerous views were one reason why he published his book in the form of a novel, a fiction asserting it had nothing to do with reality.)[14] The description of the ocean in the *Telliamed* as a process of continuous, gradual change, influenced the ideas of the famous natural historian, Georges-Louis Leclerc de Buffon in his 1749 "Theory of the Earth." Buffon described a rocky globe shaped and reshaped over time by flowing intervals of water.[15] The ocean was the ink which wrote the history of the earth upon the terrestrial globe, inscribing in the form of valleys, mountains, and chasms a story of the succession of events over a long duration: the deposition of sediments, their subsequent erosion, the creation of mountains imbedded with fossils, the retreating of the waters as these mountains grow too big to remain hidden underwater. The land covered by water was simply waiting, with time, to be exposed and then to be covered up again. Because the terms were set by a regular, natural principle of ebb and flow, erosion and deposition, submarine and dry land could be characterized using the same language for the philosopher capable of seeing all from above.

Another account that further demonstrates the understanding of the ocean in terms of the land was the *Natural History of the Sea* (1725), of the Italian naturalist Luigi Ferdinando Marsigli. The goal of this work was to find "if it existed, in the entire body of the Earth, an ordered symmetry of all the parts composing it." Marsigli hoped to show that the Biblical Flood, and other floods and cataclysms of the past, had not created random destruction and scattered ruins. Rather, it was possible to discover the overarching disposition of the Earth's geographical phenomena and understand how they corresponded to one another through "useful analysis."[16] He mentally divided the Earth into three comparable parts: higher elevations not presently touched by the waters of the ocean, liminal areas between the sea and elevated land that were sometimes afflicted by storms and floods, and the region from the shoreline into the depths of the ocean. Although the depths of the ocean were not visible to the human eye, "[t] he two first parts must inform us of how the third part is probably constructed, as [this third part] is clearly but a continuation of the two apparent ones... . Thus the bottom of the Sea is not only linked to the coastlines ... but it is indeed a very ordered continuation of them."[17] The land covered by water and hidden to the human sight looked very much like visible land, and its shape could be inferred from the corresponding dry land.

This methodological analysis, demystifying the contents and nature of the underwater realm, applied not only to geographical features but also to the plant and animal life of the ocean. Marsigli assured readers that undersea plants were products of the same divine, but also rational, plan. Every plant growing on dry land, moreover, had a corresponding version in the sea.

> Plants on earth are visible to us, but those that are under the waters of the sea are hidden to us and often remain mysterious. But, as with geography, the generation of marine vegetation perfectly corresponds to that of the land. The Basin [of the sea] was ... just as the other part of the Earth that is raised above the Horizon of the Sea, furnished by the Creator with Plants subjected to this commandment ... as it is [told] in Scripture.[18]

In fact, it was possible that there were even more plants in the sea than on land. The land was the product of the sea, which had left mineral depositions that helped plants to flourish. In the sea, plants had constant access to these minerals, and could grow without restriction; it had therefore to be the case that the undersea was "very abundant in all sorts of Plants, & even more than the Surface of the Earth."[19] All of these underwater plants would, however, be perfectly comparable to and analogous with plants of the land.

Buffon also described the richness of life in the oceans – again, always in terms of the land:

> [T]housands of habitants of different species populate the entire area [of the ocean], some covered with light scales, quickly traversing different countries, others burdened with a thick shell, heavily dragging themselves and slowly marking their path in the sand; others to whom Nature gave fins in the form of wings, using these to raise and hold themselves in the air; others, finally, to whom all movement was denied, growing and living attached to boulders; all of them find in this [watery] element their lifeblood; the bottom of the sea produces plants, mosses, and even more unique forms of vegetation in abundance; the terrain of the sea is composed of sand, gravel, often of mud, sometimes of earth, of shells, of rocks, & everywhere it looks like the earth upon which we live.[20]

The world under the sea was recognizable and continuous with the nature of the land – although it seemed to have been considered more proliferous and teeming than the dry world. Continuing the analogy between life under the sea and on land, De Maillet's *Telliamed* asserted that all life had emerged from the sea, and thus there necessarily was a visible and logical continuity from aquatic to terrestrial species. In the case of "terrestrial animals, . . . there is not a single walking, flying, or crawling one, of which the sea does not enclose related or similar species, and of which the passage from one of these elements to the other is possible, probable, and even supported by a great number of examples."[21] Some examples of these matching creatures included "sea monkeys" and "terrestrial monkeys," "sea elephants" and "terrestrial elephants," corresponding marine species for "the lion, the horse, the ox, the pig, the wolf, the camel, [and] the sheep," as well as mermaids and mermen that paralleled earthly humans.[22]

The poetic imagination of the submarine and the sea floor, that world parallel to the one above water, was perhaps best embodied by Buffon, who encapsulated the undersea in the language of earthly geography as if from above and in the blink of an eye:

> Considering . . . the bottom of the sea, we notice as many irregularities there as on the surface of the earth; we find peaks, valleys, plains, depths, boulders, all sorts of terrains; we see that islands are nothing but the summits of vast mountains. . . . There are those tempestuous regions where violent winds hasten storms, where the sea and the sky, equally agitated, collide and become confused with one another: here are the internal movements, turbulences, torrents and extraordinary unrest caused by volcanoes whose submerged mouths vomit fire from the depths of the waves. . . . Further along I see those abysses which one does not dare approach, which seem to attract and swallow ships; beyond I glimpse those vast plains, always calm and tranquil . . . finally, casting my gaze towards the extremities of the globe, I see those enormous glaciers that separate from the continents of the poles and travel, like floating mountains, all the way to the temperate regions to melt.[23]

Yet, despite the lyrical quality of this description, it remained logical because its language was continuous with that of the visible Earth. Like Marsigli and Desmarest, Buffon drew upon knowledge of the observable world to fill in the gaps produced by the invisible. He depicted this hidden world, moreover, as if he were able to float on the surface of the water, looking through it as if it were a translucent plane of glass, and visualizing its depths as if they were a vast landscape. The 'painting' that he made of the undersea world – through the lively brushstrokes of language – was intended to be understood as all the more accurate and true as all parts of it were imbued with the philosophical spirit of transforming nature into a logical machine.

Thus the study of geography merged with the study of language: both were Enlightenment sciences which sought to elucidate the process of discovering, of laying bare, the unknown in terms of the familiar and known. The rational language of nature would embody and integrate the concealed mechanics of nature with the hidden path of human reason. This ideal was epitomized in the work of Étienne de Condillac, who distinguished in his *Logic, or the initial developments of the art of thinking* (1780) between "seeing" and "understanding." Seeing the world involved identifying each of its individual components, what Condillac called "decompos[ing]" or finding "the analytic order." Understanding it meant "recompos[ing]" these isolated parts in the mind, putting them into rational and syntactical relationship with one another, and transforming them from mere physical objects to signs in a universal grammar of world and mind.[24] This language of nature was, Condillac thought, not simply a replication of the physical world, but a more accurate, truer representation of it. Through "understanding," or what he also called "mental sight," one grasped the related meaning of a series of objects in a "simultaneous," not "successive," manner. All parts of a system could thus be discerned instantaneously, and more links between objects could be discovered than had been initially thought possible. When the activities of the mind became immediate and all-encompassing, summarizing the processes of the natural world in a powerful "glance," Condillac argued that "the mind sees more than the eye can."[25] Confronted with the confusion and obscurity of the ocean, geographers and natural historians mentally and symbolically transformed it into an uninterrupted extension of the land.

When seen from above and in the mind's eye as a network of corresponding parts, the terraqueous globe became a rational, symbolic, and aesthetic experience. "Condillac's analysis is . . . a form of cartography," Wilda Anderson argues: "the map is the territory, or the territory can only be seen as such when it is converted into a map."[26] Desmarest indeed compared his comprehension of the world from above to a map: "To be convinced of [the organization and parallel structure of geographical formations across the globe], without traveling far, it suffices to look at one of our best maps."[27] The "best maps" revealed the world to be an ordered system, that became ever clearer as the observer moved upwards, abstracting local, specific observations into more general statements. That the philosophers of the mid-eighteenth century could not actually fly far above the Earth (they would have to wait until the end of the century, for experiments with hot-air balloons) did not matter, because for them, the world became more understandable and more true when empirical, sensorial data was passed through the abstracting lens of the reasoning, mathematical mind. What could not be seen with the naked eye, moreover, was invented according to the same rules and organizing principles of reason. Thus the ocean and what it contained became a rational space. So long as it could be connected to the already logical language of dry land, it fit flawlessly into a continuous argument about the geographical system of the natural world. To 'see' under its waves was the same as to climb to the top of the highest mountain, exhibiting the mental talent of purifying and crystallizing what at first seemed like a ruined, mysterious globe to its true, pure nature.

Conclusion: facing the obscurities of vision

This philosophical way of understanding the Earth and the oceans changed with the destabilizing experience of leaving dry land and losing continuity with it. As other articles in this volume reveal, the study of the effects of light and color on the perception of the undersea, as well as Romantic and psychoanalytic visions of the ocean, did not follow the rigid aesthetics and expectations of Enlightenment geographers. As explorers began to find ways to go deeper into the ocean in the following centuries, how did they explain the difficulties and distortions of perceiving underwater, which threatened to make the ocean irreconcilable with the rest of the Earth? In her article on hot-air balloons, Anderson has addressed this question in terms of the idealized space of the air, showing how the abstracted and purified view from above was upset when people actually floated above the Earth in a hot-air balloon and looked down from this height.[28] The phenomenological experience of observing what had once been logical and invisible was also captured by the Swiss naturalist Jean-André de Luc as he explored the Alps. His account of these massive mountains up close did not align with Desmarest's views of the world from above. De Luc's observations were far more full of mysteries, twists and turns, surprises, games of light and shadow, fleeting moments of pleasure: "a variety that one can barely imagine, when one has not experienced it." The landscape from a distance revealed little about the experience of actually being there. One could not enter previously inaccessible worlds such as mountains except "by winding through these labyrinths, that from far away only look like solid masses."[29] De Luc's language was playful; when describing overhangings in the mountain cliffs, he used the French word "saillies" – witticisms, pleasantries – which was also a literary, rococo term for a refined, subtle style of writing.[30] In his descriptions of the enigmatic landscape of mountains, he did not appear to be searching for an all-encompassing key, cause or explanation, nor to take the same path out as he had come in – for he admitted that he could not find that path when descending the mountain. Rather, his approach remained flexible, his mind adapting to the unpredictable.

Perhaps it was through this alternative state of perception that submarine worlds could be experienced in a closer, more direct and more sensory manner.[31] This other way of seeing might be compared to Elena Russo's discussion of a very different Enlightenment aesthetic, that of the "*Je ne sais quoi*." A text written in this style is not associated with the "timeless," pure beauty of logical, abstracted truth, Russo argues. On the contrary, it captures the "invisible and ultimately unknowable" in "a quick succession of discrete moments, each offering new and unexpected sensations; it inhabits time and history because it unfolds as a series of distinct experiences vying with one another for the beholder's attention."[32] In a description of the sea in his *Studies of Nature*, the natural historian Bernardin de Saint-Pierre reproduced this spirit of confronting the unknowable in sensory time and space. He rendered a captivating, dramatic painting of the undersea, all the while underlining his incapacity to really know it or to predict what would happen:

> If it had been in my power, I would have studied the contrasts that these innumerable families [of fish, of crabs] form upon the silt and the rocks, where their scales sparkle like the fires of dawn, with a purple and azure radiance. I would have described these Pelagian countries, covered with plants of an infinite variety of forms, that do not receive the Sun's rays except through the waters. . . . I would have represented their cliffs which rise from the bottom of the abyss like unshakeable jetties, with cavernous sides bristled with corals and papered with garlands of moving bodies of algae, seaweeds, kelps of all colors, that serve as hideaways and beds for seals and seahorses. During storms, their shadowy bases become covered with clouds of phosphoric light; and inexpressible sounds that escape their crevices call the silent legions of the inhabitants of the sea to their prey. I would have tried to infiltrate this palace of the Nereids, to unveil mysteries still unknown to men, and to observe from afar the traces of that infinite wisdom that walked upon the surges: but these laborious and delightful studies, so useful to fishing and so agreeable to natural history, are beyond the fortune and the work of a solitary [person].[33]

Saint-Pierre appealed for help in the study of the submarine world, reflecting a late-Enlightenment shift in the definition and methodologies of natural history that once belonged to Buffon and Desmarest. No longer was understanding the ocean in the power of a single, philosophical mind; it could no longer be seized and summarized from above. And yet, even though mystery beckoned from it, even though it resisted logic and control, this text remained useful to the natural history of the ocean. It seemed designed to fill the reader with inspiration to discover through a language that embraced the fear and the unknown as the very essence of what it meant to experience the underwater realm.

Notes

1 Jean le Rond D'Alembert, "Discours préliminaire," *Encyclopédie, ou dictionnaire raisonné des sciences, des arts et des métiers*, eds. Denis Diderot and Jean le Rond

D'Alembert, ARTFL *Encyclopédie Project*, ed. Robert Morrissey (Chicago: University of Chicago, n.d.) http://encyclopedie.uchicago.edu/ (accessed May 2018), Vol. 1: xv. All translations from the French are my own, unless otherwise noted.
2 D'Alembert, "Discours préliminaire," Vol. 1: i.
3 D'Alembert, "Discours préliminaire," Vol. 1: vi.
4 D'Alembert, "Discours préliminaire," Vol. 1: ix (my emphasis).
5 Nicolas Desmarest, "Géographie physique," *Encyclopédie*, Vol. 7: 613.
6 Desmarest, "Géographie physique," Vol. 7: 613.
7 For more about these authors and the context of their theories, see Norman Cohn, *Noah's Flood: The Genesis Story in Western Thought* (New Haven, CT: Yale University Press, 1996), 52–83. See also William Ashworth, *Theories of the Earth, 1644–1830: The History of a Genre* (Kansas City, MO: Linda Hall Library, 1984); Kerry V. Magruder, "Thomas Burnet, Biblical Idiom, and Seventeenth-Century Theories of the Earth," in *Nature and Scripture in the Abrahamic Religions: Up to 1700,* eds. Jitse M. van der Meer and Scott Mandelbrote, vol. 2 (Leiden: Brill, 2008), 451–90; and Maria Susana Seguin, *Science et religion dans la pensée française du XVIIIe siècle: le mythe du déluge universel* (Paris: Honoré Champion, 2001), 93–104.
8 On the rationalism of the physio-theologians, see Peter Harrison, *The Bible, Protestantism, and the Rise of Natural Science* (Cambridge, UK: Cambridge University Press, 1998), 194–98, and Magruder, "Thomas Burnet" 467–9.
9 Nicolas Desmarest, *Conjectures physico-méchaniques sur la propagation des secousses dans les tremblemens de terre: et sur la disposition des lieux qui en ont ressenti les effets* (Paris, 1756), 5.
10 "Machine (*Peinture*)," *Encyclopédie,* Vol. 9: 798.
11 Andrew Graciano, "'The Book of Nature is Open to All Men': Geology, Mining, and History in Joseph Wright's Derbyshire Landscapes," *Huntington Library Quarterly* 68: 4 (December 2005), 590–1.
12 Demarest, *Conjectures physico-méchaniques*, 5–6.
13 Benoît de Maillet, *Telliamed: ou, entretiens d'un philosophe indien avec un missionaire françois sur la diminution de la mer, la formation de la terre, l'origine de l'homme, &c.* (Amsterdam: Chez l'Honoré & Fils, 1748), xvii–xx.
14 On the *Telliamed*, see Seguin, *Science et religion dans la pensée française,* 158–66.
15 Georges-Louis de Buffon, *Histoire naturelle, générale et particulière, avec la description du Cabinet du Roy*, Vol. 1 (Paris: Imprimérie royale, 1749), 79–95. Buffon's neptunist perspective, or explanation of the Earth's history through the movement of water, would change later in his career, when he increasingly attributed the history of the globe to fire and volcanic activity. For more on Buffon's theory of the Earth in the 1750s, see Martin Rudwick, *Bursting the Limits of Time. The Reconstruction of Geohistory in the Age of Revolution* (Chicago: University of Chicago Press, 2005). On the relationship of the *Telliamed*, as well as other theories of the Earth at the time, to Buffon's work, see Jacques Roger, *Buffon. Un philosophe au jardin du roi* (Paris: Fayard, 1989), 141–4.
16 Luigi Ferdinando Marsigli, "Préface," in *Histoire physique de la mer. Ouvrage enrichi de figures dessinés d'après le naturel* (Amsterdam, 1725), n.p.
17 Marsigli, *Histoire physique de la mer*, 5–6, 9.
18 Marsigli, *Histoire physique de la mer*, 51.
19 Marsigli, *Histoire physique de la mer*, 53.
20 Buffon, *Histoire naturelle*, Vol. 1: 72–3.
21 De Maillet, *Telliamed*, 133.
22 De Maillet, *Telliamed*, 143–4. On mermaids, as well as other fantastical creatures of land and sea, see pages 151–62.
23 Buffon, *Histoire naturelle*, Vol. 1: 71–2.
24 Étienne de Condillac, *Logic*, trans. W. R. Albury (New York: Abaris Books, 1980), 78–9.

25 Condillac, *Logic*, 70–3.
26 Wilda Anderson, "Vol au-dessus d'un paradigme nouveau," in Ballons et regards d'en haut [set title from Ballons to haut in italics], eds. Michel Delon and Jean Goulemot, Cahiers de littérature française. Vol. 5 (Paris: L'Harmattan, 2007), 29.
27 Demarest, *Conjectures physico-méchaniques*, 6.
28 Anderson, "Vol au-dessus d'un paradigme nouveau," 36.
29 Jean-André de Luc, *Lettres physiques et morales, sur les montagnes et sur l'histoire de la terre et de l'homme* (En Suisse: Chez les Libraires Associés, 1778), 113–14.
30 De Luc, *Lettres physiques et morales,* 115.
31 On the transformation of the seascape into a romantic, poetic, mysterious – but also enjoyable – environment, see Alain Corbin, *The Lure of the Sea: The Discovery of the Seaside in the Western World, 1750–1840,* trans. Jocelyn Phelps (Berkeley: University of California Press, 1994).
32 Elena Russo, *Styles of Enlightenment: Taste, Politics, and Authorship in Eighteenth-Century France* (Baltimore: Johns Hopkins University Press, 2007), 158–9.
33 Bernardin de Saint-Pierre, *Études de la nature*, Vol. 3 (Bâle, 1795), 498–9.

4 Understanding the loss of colour

Jonathan Lamb

The specific loss of colour I want to discuss is taking place along the Northeastern coast of Australia where the Great Barrier Reef, consisting of the largest known formation of corals in the world, is rapidly degrading and dying. The human contribution to this disaster is direct – fertiliser and silt run-off from the sugarcane plantations on the Queensland shore, and overfishing. But the worst effects are indirect – the warming of the sea owing to climate change and increasing ocean acidification as carbon dioxide is converted into carbonic acid. Nitrogenous run-off and coastal sediment harms coral and encourages predators such as the crown of thorns starfish, meanwhile the absence of fish stimulates the growth of algae leaving less space for coral to grow; acidification dissolves the calcareous structure of the coral habitat; and ocean warming reduces the food supply on which coral depends, provided by its commensal partners called zooxanthellae. When it gets too warm these microscopic plants induce oxidative stress in their hosts by mixing free radicals with their carbohydrates; the corals expel them, then turn white, for the zooxanthellae are the source of all the glorious colours of the reef, and should the bleaching occur more than once, the coral polyps will die.

So the most immediate and dramatic feature of the material loss of the Great Barrier Reef is the disappearance of colour. In December 2016 *The New York Times* carried a précis of an Australian Research Council report to the effect that in the course of that year a large extent of the Reef's northern areas had suffered the worst coral bleaching and die-off ever recorded.[1] Into the midst of this crisis the Queensland Government had already thrust its own contribution, backing a plan of the Adani Group of Gujerat to develop the Carmichael Project. This will open the vast coal reserves of the Galilee Basin in central Queensland for mining, and connect the Basin with two deepwater coal terminals at Point Abbott and Hay Point via a 300 mile rail-link. None of this is yet built and so far loans and subsidies are not forthcoming, and the minority Labour government of Queensland, which came to power on a wave of public indignation when the plan was first mooted, is opposed to it.[2] If Adani succeeds it will load sufficient coal into VLBCs for transport to India to produce 700m tonnes of carbon dioxide over the fifty years' life of the operation.[3] The long-term damage caused to reef systems by acidification and warming will straight away be compounded by drift of coal-dust over the reef and the physical damage done to the coral by the dumping of 5m tonnes of spoil from the deepening and widening of the ship-channels at sea, not to mention the constant passage of such large ships through the reef. These plans were suspended while Australia was trying to avoid having the Reef listed as an endangered World Heritage Site in 2014–15, but now powerful voices are calling for their resumption, including that of the Australian Prime Minister, Malcolm Turnbull.

While on their own Australians cannot contain the effects of ocean acidification, here is an example of intended damage to one of the biggest and ecologically the most vital of all coral reefs in the world. The sole motive alleged is the unemployment among miners and workers in associated industries, since global sea-borne coal traffic currently needs no supplement. Profit to be gained from the venture is uncertain (four of Australia's biggest banks have refused to fund it) and it would be more than offset by the damage done to the tourist industry of Queensland. That the project was shelved in the face of public opposition and has now been renewed is a symptom of the invincible commitment to the welfare of extractive industries in Australia, shared by Republicans in America whose support for an unregulated coal industry in Kentucky, West Virginia and other coalmining states in America has been given a big boost by the executive orders of President Trump, not to mention the recent decision to shrink national parks in Utah and Alaska in favour of oil and gas drilling.

The assumption common to the coal lobbies in both countries is that the world is an infinite common, perpetually exploitable, and that climate change is a hoax. But I believe there might have been a darker impulse for this folly, running deeper in Western epistemology and the postcolonial psyche than mere greed or caprice. The blazon of these people is black for coal and white for dead coral. Goethe provides half of the bleak commentary on it in his *Theory of Colours* when he says, "Everything living tends to colour – to local, specific colour. . . . Everything in which life is extinct approximates to white, to the abstract, the general state."[4] Ruskin supplies the rest when he tells his reader:

> Take a wider view of nature, and compare generally rainbows, sunrises and corals with fungi and fogs, and consider for a little while what sort of a world it would be if all flowers were grey, all leaves black, and the sky brown and you will feel how the question stands between the colourists and the chiaroscurists, which of them have nature and life on their side, and which have sin and death.[5]

Ever since Plato decreed that the truth of things lay in intelligible forms that were inaccessible to the senses, the value of the empirical evidence of the senses has been losing ground to mathematical arrangements of information that we are told represent the only reliable knowledge, possibly even the truth itself. In a strange experiment upon himself Descartes imagined that all his senses were inoperative, and all the ideas derived from them were forgotten. What he then began to do was think, in the dark, and what he thought was something like intelligible forms. Locke would later call these primary qualities, what cognitive science these days (taking the mind for a machine) conceives as "abstract, symbolically defined input-output mappings." That phrase belongs to Tim Ingold, who persuasively argues that cognitive science in its various manifestations, which would include digital humanities and the technique of "distant reading," is firmly grounded in

> the Cartesian ontology that is basic to the entire project of cognitive science – an ontology that divorces the activity of the mind from the body in the world. . . . What [its disciples] fail to recognize is that the processing loops that yield intelligent action are not confined to some interior source of the mind, confined within the skull, but freely penetrate both the body and its environment.

As far as the eye was concerned, Descartes thought it did little more than a blind man's stick in discovering the geometry of objects. Its job was to send signals the mind could convert into a semiotic pattern of vision, which can never resemble the object because vision has turned into language, and light into the darkness of Cartesian thinking. This marks the difference between chiaroscurists and colourists.[6]

In the vicinity of the Reef the chiaroscurists don't have it all their own way. *The New York Times* reported that a team of NASA scientists based at Heron Island, situated in the southern part of the Reef not very far from the planned terminals at Gladstone, were using a specialist camera in the belly of a Gulfstream jet to determine infinitesimal alterations in the widths of the spectral colour-bands in order to judge how well the coral was coping with its latest stresses.[7] If you set the priorities of the NASA scientists at Heron Island against those of Tony Abbott, the previous prime minister, and his successor you see that colour is what divides them. On the one hand is the extraordinary care taken over the minute discrimination of spectra, so delicate that new names will have to be invented for the finest of the intermediate shades, all of which will be signs of life; and on the other is a pattern of decisions leading to a world where there will be no colour at all, just the dismal chiaroscuro of bleached coral and black silt. Charlie Veron, former chief scientist at the Australian Institute of Marine Science, is convinced this colourless world is closer than we think. Half the Reef has gone in the last thirty years; and he predicts at the end of another thirty it will be totally dead.

In his *Notebooks* Coleridge observed, "It is by a negation and voluntary Act of no-thinking that we think of earth, air, water etc. as dead."[8] So where did this deadly no-thinking come from and how is it deployed? To begin with it wasn't so much negation as resistance to sensation; or at least a profound skepticism about what sensation told us of the world. Descartes followed Plato in his suspicion of any form of represented (i.e. empirical) knowledge, as opposed to the immediate and demonstrable truths of geometry: so the senses, especially the eye, deluded us with sensory information that didn't really belong to the things to which we attached it. Colour as a property of an object was a delusion. In his *Treatise on Light* he wanted his reader to entertain the possibility of "a difference between the sensation that we have of it . . . and what it is in the flame or in the Sun that we term 'light.'"[9] He concluded, "I cannot accept the distinction the Philosophes make between true Colours and others which are only false or apparent." He saw all colours as merely apparent. The yellow of the sunflower doesn't reside in its petals but in the signals passing through the brain of its observer. The true properties of all things he listed as abstractions: bulk, figure, number and motion. Following the Cartesian line on sensation, Locke wrote,

> Ideas or Perceptions . . . we *may not* think . . . are exactly the Images and Resemblances of something inherent in the subject; most of those of Sensation being in the Mind no more the likeness of something existing without us, than the Names, that stand for them, are the likeness of our Ideas.[10]

So he concluded, "The cause of Colour in external Objects may be only a privation," that is, the sign of nothing.[11] Everyone contributing to this line of thought, from Descartes to Berkeley and from Locke to Thomas Reid, will insist on the analogy between cognition and language. The impression of a thing forms the sign of an idea in the same way that a word, by convention and agreement, stands for a determinate meaning. Sentience is merely the portal to a system of semiotics, and once a sensation has turned into an idea it cannot return to the status of a sensation. Locke concluded, "There is nothing like our Ideas, existing in the Bodies themselves . . . and what is Sweet, Blue, or Warm in Idea, is but the certain Bulk, Figure, and Motion of the insensible Parts in the Bodies themselves, which we call so."[12]

The implications of Locke's line of thought struck directly at the aesthetics developed by Joseph Addison in his *Spectator* papers on the pleasures of imagination. He was all for the reality of colour, ("The Eye takes most Delight in Colours . . . [in] the different stains of Light that shew themselves in the rising and setting of the sun . . . we find the Poets . . . borrowing more of their epithets from Colours than from any other Topic").[13] Having spent some time explaining the delicious complexities of the camera obscura, where the colours, shapes and movements of a real landscape are projected so faithfully on to a screen it is impossible to tell the original from the copy, he declares, "What a rough unsightly Sketch of Nature should we be entertained with, did all her Colouring disappear, and the several Distinctions of Light and Shade vanish?" It comes to him therefore as a grievous concession when he invokes the consensus now largely surrounding the great modern discovery in natural philosophy: "Namely that Light and Colours, as apprehended by the Imagination, are only Ideas in the Mind, and not Qualities that have any Existence in Matter."[14] It leaves Addison adrift, alone and palely loitering like a knight in a chromatic romance "who sees beautiful Castles, Woods and Meadows . . . but upon the finishing of some secret Spell, the fantastic Scene breaks up, and the disconsolate Knight finds himself on a barren Heath, or in a solitary Desart." He allows himself to hope that the departing soul might be cherished by ideas of colours deriving from some different source, but it is an uncertain consolation. Addison was speaking on behalf of those who take aesthetic pleasure (and even disgust) seriously and who, like Hume, were prepared to stand up for the reality of the resemblance between sensations and ideas. Hume it was who entirely set aside the great discovery that there was none when he maintained in his *Treatise* that, "That idea of red, which we form in the dark, and that impression, which strikes our eyes in the sun-shine, differ only in degree, not in nature."[15]

On whose behalf was Locke arguing? I'm tempted to say: those who were not frightened of nothing. After all, in his second meditation Descartes had repudiated all of empirical experience as the trick of vile enchanter, leaving him (as he said) feeling like a god, not at all like a forlorn knight at arms. His constituency embraced all those scientists, philosophers and discoverers who believed it was possible to discover truth in the dark, or at least without the aid of the five senses. A dungeon for Addison had been a lightless place where the imagination could supply all manner of consolations, just like the camera obscura. But when Descartes hypothesized that every sensory input was a gross fiction passed upon the mind by a malign djinn, he placed himself in an impenetrable darkness that provided him with the certainty of ideas deriving from some other source than sensation. Absence of light yielded him the geometrical truths of figure and the mathematical truths of number, and to his descendants he left the legacy of calculus, the truths of linear motion, whose achievements Berkeley thought were an elaborate adventure with nothing: "That is, they consider Quantities infinitely less than the least discernible Quantity ... so there are Differences ... in an infinite Progression towards nothing, which you still approach, and never arrive at."[16] The line of velocity, he said, "entirely stripped of all other sensible qualities and circumstances ... is neither black nor white, nor red, nor hath any colour at all, or any tangible quality whatever."[17]

As opposed to the empiricists of the Royal Society, such as Sprat, Boyle and Hooke, whose curiosity took the form of the hypothesis, "What if what I don't know, I did," the Cartesians supposed the opposite: "What if what I do know, I didn't." Blake's picture of Newton shows the inventor of fluxions calculating conic sections while seated on a bank of coral, oblivious to its beauty (Figure 4.1, see plate section). Although this makes a powerful point about the mathematisation of knowledge, it isn't quite fair to Newton, who argued for a vibrational link between colour and the optic nerve, and between the optic nerve and the sensorium: that is, an unbroken physical link between what happens in the eye and then takes place in the mind.[18] Associationists in the eighteenth century adapted Newton for a strong defence of empiricism, beginning with David Hartley. Erasmus Darwin, the heir along with Joseph Priestley of this line of thinking, simply asserted, "The animal activity of the retina constitutes vision."[19]

When Locke proposed to himself the problem of the nature of darkness, it was not to imagine that he might be able by some means or other to cast a light on it. What he wished to see was darkness itself. He entertained the ingenious idea that privation was essential to any idea of fullness, as shadow is to light, and chiaroscuro is to figure. He demands,

> Whether the shadow of a Man, though it consists of nothing but the absence of Light (and the more the absence of Light is, the more discernible is the shadow) does not, when Man looks on it, cause as clear and positive an Idea in his mind, as a Man himself, though covered over with clear Sun-shine?[20]

This strangely emphatic contrast between two men, the one defined by the absence of light and the other totally irradiated, anticipates Addison's own miserable speculation that the enchantment of the living eye disporting itself in light will be followed after death by the departing soul's bleak vision of a world reduced to a cinder. This is what Locke goes on to say, "Indeed, we have negative Names, which stand not directly for positive Ideas, but for their absence, such as *Insipid, silence, nihil,* etc. which Words denote positive Ideas; v.g. *Tast, Sound, Being,* with a signification of their absence."[21] It sounds as if he might be making the same claim for the dialectics of chiaroscuro mounted by Ruskin in *Modern Painters* when he says that Turner attains brilliancy of light by "clear and exquisite drawing of the shadows ... shadows sharp and clear."[22] But that is not what Locke means, for he concludes his estimate of the positive product of privation with this remarkable statement: "And thus one may truly be said to see Darkness. For suppose a hole perfectly dark, from whence no light is reflected, 'tis certain one may see the Figure of it, or it may be painted."[23] This is what Eugene Thacker calls "retinal pessimism" which he glosses as "the showing of nothing ... this 'nothing-to-see.'"[24]

Anyone on the empiricist side of this debate would say with Aristotle that substantial darkness is the total absence of light and therefore the loss of visibility necessary for any action of the eye. Boyle was interested in the adjacency and mixture of blindness and sight, a topic anticipating the Molyneux debate about how people suddenly cured of blindness would see things. His fascination with illnesses that affected sensory organs only to leave them afterwards more acutely receptive to impressions included two cases of challenged sight that overcame the totality of darkness. He mentions a man whose sight was so fine after an inflammation of the eye that he could see colours in the dark. But his premier instance is the blind Dutchman John Vermaasen who could distinguish colors by touch. Both examples begin to rehearse Addison's claim that sight is "a more diffusive kind of Touch," and to introduce possibilities of a synaesthetic relation of the tactile and the visible that Berkeley explores in his *New Theory of Vision*. They also endorse emphatically Ruskin's insight into the privation of light as useful only insofar as it is a stimulus to sight, and provide an illustration of Goethe's dictum that "colour itself is a degree of darkness," i.e. a phenomenon less than white.[25]

Now I want to draw a parallel between Locke's method of a privative knowledge and the discoveries made by Charles Darwin when studying the growth and death of coral reefs. He espoused the theory that the subsidence of the ocean floor was the motor of the upward growth of corals, but he did not know this, so he had the same recourse as others lost in wonder in the presence of corals, namely imagination. "Let us in imagination place within one of the subsiding areas, an island . . ." says Darwin, sounding like Prospero as he begins his conjecture of how a coral reef is formed.[26] The slow subsidence of the ocean floor, he surmised, is an imperceptible geological descent answered by a biological ascent of coral whose rate of progress is impossible to ascertain: "We have no precise evidence on this point, and comparatively little concern with it."[27] If that constitutes the vertical axis of the coralline enigma, then its horizontal coordinate is supplied by the force of the open ocean as its swells break on the coral that has reached the surface, either eroding it and sweeping the detritus into an accumulation that will become an atoll, or replenishing the corals on the very edge of the windward part of the reef so that they can extend sideways the platform on which it stands. Destruction is not only simultaneous with construction, it actually participates in it, for as the sea abrades the coral into a cement that forms the platform of the advancing live coral, so the calcareous parts of the living coral consumed by fish, sea-cucumbers and those ghastly pink worms that so revolted Robert Louis Stevenson when he landed in South Seas, are cast as fine coral silt which provides more cement for the atoll. Although Darwin generally depicts this as a struggle between two antagonistic principles, "the two nicely balanced powers of land and water," and more specifically, "the vital energies of the corals versus the mechanical power of the waves," he concedes that the contest is nowhere near so symmetrical: "The relations which determine the formation of reefs on any shore, by the vigorous growth of the efficient kinds of coral, must be very complex, and with our imperfect knowledge quite inexplicable."[28]

It is not then any false modesty on Darwin's part that leads him to confess that while aspects of the formation and ruin of coral over time might be conjectured with a fair degree of probability, the system as a whole is inexplicable, leaving our knowledge of it imperfect. The same cognitive deficit that obliged Newton to suppose a sea-bottom coloured red after he heard of the changes of light that took place in Edmond Halley's diving bell, leaves Darwin relying on imagination. Yet while he is quite willing to leave what does not concern him in a state of inexplicability, he believes he is close to knowing *something* substantive about the relation of subsidence to coral reefs, having recruited not only imagination for his researches but also a vast quantity of empirical testimony. The two words "probable" and "inexplicable" glance in opposite directions in Darwin's *The Structure and Distribution of Coral Reefs*, for the former heralds the arrival at some sort of evidence, solid and real, for an hypothesis that will be confirmed in the future, and to which we will have been conducted by probable reasoning. "Inexplicable" on the other hand refers to a life-activity of coral so complex that it cannot be disintricated. Darwin drew his probable conclusions from drowned coral reefs in the Indian Ocean, the Great Chagos Bank and the Keeling Atoll, "an irregular field of branching corals, still upright, but entirely dead ... of a brown colour."[29] Privation of coralline life served like Locke's privation of light to give Darwin's researches a certain strong figure or frame that otherwise they would have lacked. He concluded, "It is, then, not improbable that the corals should sometimes perish ... As by our theory ... there is nothing improbable in the death ... of corals."[30]

Darwin's double negative reminds me of Hobbes who, in his *Elements of Philosophy*, observed, "the word nothing is not unuseful," adding,

> Positives were before negatives ... When the name of *white* was imposed upon certain things, and afterwards upon other things, the names of black, blue, transparent, etc. the infinite dissilimitudes of these with white could not be comprehended in any one name, save that which had in it the negation of white, that is to say, the name *not-white*, or some other equivalent to it.... And by these negative names we take notice ourselves and signify to others what we have not thought of.[31]

You see how Hobbes proceeds in a direction the opposite of Locke's and Charles Darwin's, using a negative to reserve a place for the subtlety of future positives instead of obliterating it. In fact, Hobbes anticipates the rationale of the deployment of the NASA Portable Remote Imaging Spectrometer which will detect bands of colour of which we have been hitherto unaware and for which we have as yet no names; and at the same time he rehearses the actual history of the sensations and names of colours, such as green and blue, which had very uncertain designations in the ancient world not because the sentient principle of people then was any less fine than ours, but because occasions for the discrimination were not as likely to be recorded in written language or held to a common standard. The anthropologists W.H.R. Rivers and Franz Boas were both intrigued by the differences in the sense perception of colours between themselves and the communities they visited. Rivers was particularly interested in blue, and would show the people a blue garment and then ask them to identify the tint on a set of glass slides of different blues, traversing the spectrum from rich and warm to thin and cold, but seldom did his sense of the common blue specimen match the slide they chose to represent it. Hume supposed a similar experiment, having a man ignorant of a certain tint of blue be shown the whole spectrum of blues except that one he has never seen, which would appear a blank. Would he be able to imagine it? Hume asks, and answers that he would, indicating that ideas are not only derived from sensations but capable of generating them too.[32] The cultural implications of this insight are extensive, suggesting that the discrimination of a tint is not referable to an archetype or form, but to associations of impressions and ideas that may be common to a community or a place, but not to a criterion. "Not-white," that is to say, has all sorts of possibilities, and they are all coloured for the reason that ideas and sensations act and react upon one another, and are capable of supplying what is missing in the spectrum from the traces of adjacent colours already experienced.

The Great Barrier Reef offers two important 19th-century examples of the negative and positive possibilities of privation. In 1847 Joseph Jukes, naturalist on the *Fly*, landed on One Tree Island, a cay at the centre of what is still one of the most beautiful fringing reefs in the entire length of the Great Barrier system. Jukes said, "it consisted of compact, tough, but rather soft and spongy rock ... rather fine grained, and only here and there exhibited any organic structure or remains." He went on:

> There were no signs of living coral except a few stunted specimens in some of the deeper holes of the reef, where also were some dead masses still standing in the position of growth. The whole was very different from any preconceived notions of a coral reef, and I erroneously imagined it must be an exception to their general character; it looked simply like a half-drowned mass of dirty brown sandstone.[33]

William Saville-Kent, whose monumental study, *The Great Barrier Reef of Australia* (1893), profusely illustrated with photographs and chromolithographs, offered the British public its first clue to the remarkable extent and vividness of coral structures, was puzzled by Jukes's reaction. He rightly suspected that Jukes studied the island at the wrong time of the tide, or possibly after a storm, and was vainly looking for coral on the platforms, or flats as Darwin calls them, where very little was ever to be found. Jukes's impression of the reef, however, is frequently mirrored in Saville-Kent's own black and white photographs, showing storm-stranded boulders of dead madrepore in a prospect of rubble, a dismal show of colourless ruin very remote from the brilliance of his chromolithographs (Figure 4.2; Figure 4.3, see plate section).

But at least Saville-Kent is (like Addison) aware of beauty in the midst of desolation, and very alert to colour as a sign of life. Like Darwin, Jukes was happier with something definite, or at least not improbably linked to the extinction of life and light, instead of phenomena difficult to explicate without proper media. He lacked the equipment to study and record coral underwater. Saville-Kent on the other hand was a skilled photographer and had various devices for taking underwater photographs, as well as a fine hand for the design and colouring of the chromolithographs. The chiaroscurist in Saville-Kent was impelled into colourism in order to supply what was missing from images in black and white, like his still life of shells (Figure 4.4), whereas it is not improbable the view of coral death removed for Jukes and Darwin the colourful but awkward inexplicability of reef-life.

Colourism provides a useful corrective to Kant's contemptuous dismissal of any notion that the immediate empirical evidence of a colour such as green can ever aspire to the standard of the beautiful, owing to the discontinuity or ambiguity of the sensation of it. It is for him a question of *form*, of an abstract idea fetched from identical examples:

Figure 4.2 William Saville-Kent, "Hurricane-Stranded Corals, Fringing Reef, Port Denison," in *The Great Barrier Reef of Australia* (London: W.H. Allen, 1893), plate XXXI.

Figure 4.4 William Saville-Kent, "Barrier Reef Shells on Sponge Table," in *The Great Barrier Reef of Australia* (London: W.H. Allen, 1893), plate XLII.

> The purity of a simple mode of sensation means that its uniformity is not disturbed or broken by any foreign sensation.... For this reason all simple colours are regarded as beautiful so far as pure. Composite colours have not this advantage, because, not being simple, there is no standard for estimating whether they should be called pure or impure.[34]

Of course Kant had no particular investment in the chromatic aspect of such purity, for he treats it as a valuable sensation only insofar as it will metamorphose easily into an idea, leading Ruskin to exclaim, "Half the degradation of art in modern times has been owing to endeavours, much fostered by the metaphysical Germans, to see things without colour, as if colour were a vulgar thing."[35] Of all critics of the fine arts Ruskin seems to have been most sensitive to the composite qualities of colours and of the mobility of the situations in which they are to be found. He it was who recurred to Hobbes's refinement of the sensation of non-white while commenting on the difficulty of translating the effects of light to white paper, whose dullness is heightened by the brightness of blue sky and clouds. And in talking of the effects of a sunset on the surface of a working sea he emphasizes how far short this leaves the observer from expressing its inexplicable and evanescent complexity (Figure 4.5, see plate section):

> The whole sky from the zenith to the horizon becomes one molten mantling sea of colour and fire; every black bar turns into massy gold, every ripple and wave into unsullied shadowless crimson, and purple, and scarlet, and colours for which there are no words in language, and no ideas in the mind—things which can only be conceived while they are visible.[36]

Ford Madox Ford complained that in a single page praising one of Turner's pictures, Ruskin used the word "gold" five times and that in only one instance did the colour alluded to even faintly compare with his own idea of gold. It is a good example of creative no-thinking on Ruskin's part, analogous to Hobbes's reasons for finding the word nothing not unuseful, where privation or words or the double negative makes a place for the existence of a colour not yet known (not-white) by denying its opposite. There is none of this tension in Locke's discussion of names for negatives such as silence, which do not imply so much as declare the absence they denote, like his figure of darkness itself. I wonder what he would have said of the implications of words like "innocent" or "immaculate" where the intuition of purity floats, like Ruskin's "gold" or Hobbes's "non-white," in a sea of possible ways of not being nocent or maculate. Hooke was marveling at this kind of negative when he said of a point in calculus, the index of a position nothing can really occupy, "No Sense, or Imagination, or Fant'cy, can reach it, nor words describe it, but by a Negative, to tell you what it is not."[37]

My point of departure was and is the Reef, a phenomenon full of colour that is daily being lost as it shrinks and bleaches. In trying to understand this material loss, I have taken to amateur plein air water-colourism, using a coalition of hand, pigment and eye to represent as immediately as I can the excitement of seeing something with no ideas, machines or numbers intervening: or as Ruskin puts it, rendering things which can only be conceived while they are visible. In August 2016 I went to Heron Island and One Tree Island with a paintbox, a sketchpad and a large supply of carrots, and I took myself like the Fisher King to the margin of the land and assembled some fragments to store against its ruin (Figure 4.6, see plate section).

Notes

1 Michelle Innis, "Great Barrier Reef Facing Multiple Threats, Report Says," *New York Times*, December 2, 2016, A6.
2 "Pier Pressure," *The Economist*, November 25, 2017, 38.
3 Jamie Smyth and Kiran Stacey, "Adani Pushes Ahead with Australia Coal Mine Despite Protests," *Financial Times*, June 7, 2017, 17.
4 Johann Wolfgang van Goethe, *Goethe's Theory of Colours*, trans. Charles Lock Eastlake (London: John Murray, 1810; reprinted Cambridge: Cambridge University Press, 2014), 234.
5 John Ruskin, *Modern Painters*, ed. David Barrie (New York: Knopf, 1987), 455.
6 *Tim Ingold, The Perception of the Environment: Essays on livelihood, dwelling and skill* (London and New York: Routledge, 2000), 165, 255-7.
7 Michelle Innis, "Inspecting Reefs From 28,000 Feet Above," *New York Times*, September 27, 2016, A4.
8 Samuel Taylor Coleridge, *Notebooks*, ed. Kathleen Coburn, vol. 3 (London: Routledge, 2002), 3632 L.100 f44.
9 René Descartes, *Treatise on Light* in *The World and other Writings*, trans. and ed. Stephen Gaukroger (Cambridge: Cambridge University Press, 1998), 3.
10 John Locke, *An Essay Concerning Human Understanding*, ed. Peter H. Nidditch (Oxford: Clarendon Press, 1979), 134.
11 Ibid., 133.
12 Ibid., 137.
13 Joseph Addison, "No. 412," in *The Spectator*, ed. G. Gregory Smith (London: Dent, 1907).
14 Joseph Addison, "No. 413," in *The Spectator*, ed. G. Gregory Smith (London: Dent, 1907), 65.
15 David Hume, *A Treatise of Human Nature*, eds. L.A. Selby-Bigge and P.H. Nidditch (Oxford: Clarendon Press, 2009), 3.
16 George Berkeley, *Essay towards a New Theory of Vision*, (London: Dent, 1972), 11.
17 Ibid., 69.
18 Isaac Newton, *Opticks* (New York: Prometheus Books, 2003), 353.
19 Erasmus Darwin, *Zoonomia*, (London: J. Johnson, 1801), 1.19.
20 Locke, *An Essay Concerning Human Understanding*, 133.
21 Ibid.
22 Ruskin, *Modern Painters*, 79–80.
23 Locke, *An Essay Concerning Human Understanding*, 133-4.
24 Eugene Thacker, "Black on Black," in *The Public Domain Review, Selected Essays, Vol. III*, ed. Adam Green (Cambridge: Public Domain Review, 2016), 25, 28.
25 Goethe, *Goethe's Theory of Colours*, 31.
26 Charles Darwin, *The Structure and Distribution of Coral Reefs* (London: Ward, Lock & Co., 1910), 75.
27 Ibid., 61.
28 Ibid., 24, 53, 52.
29 Ibid., 22.

30 Ibid., 81.
31 Thomas Hobbes, *Elements of Philosophy,* in *The Works of Thomas Hobbes*, ed. Sir William Molesworth (London: John Bohn, 1839), 1.18.
32 Hume, *A Treatise of Human Nature*, 6.
33 William Saville-Kent, *The Great Barrier Reef of Australia* (London: W.H. Allen, 1893; reprinted Provo, UT: Repressed Publishing, 2015), 106.
34 Immanuel Kant, *The Critique of Judgment*, trans. James Creed Meredith (Oxford: Clarendon Press, 1952), 66.
35 Ruskin, *Modern Painters*, 455.
36 Ibid., 73.
37 Robert Hooke, *Posthumous Works*, ed. Richard Waller (London: Samuel Smith and Benjamin Walford, 1705), 65–6.

5 Hydromania

The social history and literary significance of Romantic swimming

Robin Jarvis

"Hydromania" was the name of one of three prints of Lyme Regis produced by the caricaturist George Cruikshank, based on sketches by Captain Frederick Marryat, in 1819 (Figure 5.1).[1] It depicts in mildly titillating detail assorted women, all completely naked, emerging from bathing machines and enjoying a dip in the shallow water, both with and without the help of professional bathing-women; they are the object of voyeuristic scrutiny by local fishermen and a gentleman on the far right who is making use of a telescope. The subtitle, "A Touch of the Sub-Lyme and Beautiful," both puns on the name of this fashionable seaside resort and playfully references Edmund Burke's influential 1757 treatise on landscape aesthetics. It might even be argued that, since Burke had drawn explicit analogies between the attributes of a beautiful natural scene and the contours of a woman's body, Cruikshank's print turns Burke's rhetoric back on itself with considerable satiric force.

Fast forward nearly two hundred years to 3 May 2010 and a not totally dissimilar scene on a beach in Turkey. Hundreds of very serious-looking swimmers – mostly British, predominantly male, all clad in wetsuits – congregated on the foreshore at Eceabat to mark the anniversary of the most famous swim undertaken by the most famous swimmer of the Romantic era. Over one hundred hardy individuals braved strong currents, swarms of jellyfish, and a chilly water temperature of thirteen degrees to emulate Lord Byron's feat of crossing the narrow channel of the Hellespont between European and Asiatic Turkey – a performance of which he was inordinately proud and which he undertook to prove the practicality of the exploits attributed to the mythical Greek youth, Leander, who was said to have swum the Hellespont on a nightly basis to visit his lover on the opposite shore. One of Byron's descendants was among the participants in the bicentennial swim; of the others, according to a *Guardian* journalist, some "were experienced channel swimmers, others just recreational athletes, but all were united by a love of the water."[2]

Here, then, are two very different expressions of hydromania, sea-bathing and endurance swimming – the one associated with the pursuit of health, the other with athletic accomplishment; the one ripe with eroticism, the other

Figure 5.1 George Cruikshank after a sketch by Captain *Marryat, Hydromania or a Touch of the Sub-Lyme and Beautiful!,* 1819. Reproduced by kind permission of Lyme Regis Museum.

tinged with egotistic self-display. Together, they have much to say about the phenomenon that I address in this essay, namely Romantic swimming. In what follows I shall explore the origins and context of the Romantic generation's passion for swimming, consider some of the literary meanings and uses of swimming in writing of that period, and look in particular at the way swimming features in the lives and works of three notable hydromaniacs – Coleridge, Byron, and Keats.

It is worth enquiring into the origins of that curious word, "hydromania." According to the *OED*, the first recorded use of "hydromania," defined simply as "a mania or craze for water," occurs in a letter written by Robert Southey in 1793. In describing three weeks' holiday spent touring the south of England, Southey mentions some extravagant indoor water features he has seen at a gentleman's estate in Oxfordshire. These have led him to conclude that "the hydromania is almost as bad as the hydrophobia."[3] Whether the word is actually Southey's coinage cannot be known for sure, but it seems clear that "hydromania" is of fairly recent derivation, and, as is so often the case, is formed as a punning antonym to an already existing word, "hydrophobia." The latter is recorded in purely medical usage – an "an aversion to water or other liquids, and difficulty in swallowing them" – as far back as the sixteenth century, and in a more general or colloquial sense of a "dread or horror of water" is first found in Laurence Sterne's novel, *Tristram Shandy*, in 1760. Sometime towards the end of the eighteenth century, therefore, a craze for water develops in contradiction of, or in tension with, a pre-existing horror of water. When Coleridge, who delighted in word-play, begins a letter in 1821 with the sentence, "Hydromaniâ Hydrophobia: from Water-lust comes Water-dread"[4] (he uses both terms as metaphors for mental states or activity) he has things back to front: historically, water-lust succeeded water-dread. Of course, the usages I have quoted are limited and specific, but I believe the narrative they encapsulate mirrors the emergence of other forms of Romantic water-lust in the late eighteenth century.

One of the initial spurs to my interest in this topic was a comment made by the historian Keith Thomas in his well-known study, *Man and the Natural World.* Thomas argues that in early modern England there was widespread anxiety about "any form of behaviour which threatened to transgress the fragile boundaries between man and the animal creation"; swimming was suspect, Thomas notes, because "it was essentially a non-human method of progression."[5] This philosophical aversion did not go entirely unopposed, as I shall show. Nevertheless, the basic story of swimming, at least in a British context, is one that starts in early modern times when this activity was the object of mistrust or distaste – despised by educators (as Michael West has pointed out) "as a mechanical rather than a liberal art," frowned upon by certain religious authorities as a violation of natural hierarchy, and advocated chiefly as a tool of self-preservation or for its utility in warfare.[6] These negative attitudes were not decisively challenged until the eighteenth century, when the growing popularity of sea-bathing and the elaboration of a powerful new discourse concerning the health benefits of sea water, administered both externally and internally, began to shift public opinion. Bathing, of course, is not the same thing as swimming: one can bathe without being able to swim, and some guides to sea-bathing advise taking only a single "plunge" in water of a safe depth. By the end of the century, though, recreational swimming was very much in vogue and, along with walking and mountain-climbing, had become one of those outdoor, very physical pleasures that seem to have been just as important to the Romantic generation as their intellectual or spiritual pursuits.[7]

The rise of sea bathing in the eighteenth century is too well known to need more than a quick summary here. Sir John Floyer's *History of Cold Bathing*, first published in 1715, which assembled masses of evidence from ancient texts, contemporary medical sources, and personal experimentation to prove the value of immersion in cold water to a wide variety of ailments, was one of the weightier advocates of a health argument that was pushed relentlessly to Georgian readers. At the other end of the century Thomas Reid's *Directions for Warm and Cold Sea-Bathing* (1798) continued to press the salutary effects of the sudden shock of cold water rousing the energy of the system and producing a "general warm glow."[8] Successive chapters dealt with the use of sea-bathing to treat scrofula, eruptions, fevers, gout, rheumatism, inflammatory complaints, chlorosis, and oedemas. Rather more worryingly, Reid also recommended drinking half a pint of sea water a day in the belief that this had a strong and beneficial purgative effect. While this part of his advice would probably not find many takers nowadays, Susie Parr has interestingly pointed to several modern studies apparently confirming the beneficial effects of regular cold-water bathing, including strengthening the immune system, increasing metabolic rate, decreasing blood pressure and cholesterol level, and boosting fertility.[9]

Intense marketing of medicinal sea-bathing was responsible for the rapid development of seaside resorts across the country. The patronage of George

III, who famously plunged in the sea at Weymouth in 1789 to the accompaniment of a chamber orchestra, merely gave royal approval to a well-established economic and social trend. Jane Austen, who was herself an enthusiastic bather at Lyme Regis, made the reckless ambitions of seaside entrepreneurs the focus of her last novel, *Sanditon*, left unfinished on her death in 1817. The eponymous resort, an unprepossessing village on the south coast, is the pet project of a Mr Parker, who is intent on attracting a fashionable crowd with the twin attractions of "saline air and immersion":

> no person (however upheld for the present by fortuitous aids of exercise and spirits in a semblance of health) could be really in a state of secure and permanent health without spending at least six weeks by the sea every year. – The sea air and sea bathing together were nearly infallible, one or the other of them being a match for every disorder, of the stomach, the lungs or the blood; they were anti-spasmodic, anti-pulmonary, anti-sceptic [*sic*], anti-bilious and anti-rheumatic.[10]

As the characteristic ironic tone indicates, Austen had little time for the aggressive marketing of sea air and sea water – a point underlined by the fact that this fragmentary text is populated almost entirely by hypochondriacs. Nevertheless, she could not get enough herself of the sea water at Lyme, writing to her sister Cassandra on one occasion that "The Bathing was so delightful this morning . . . that I believe I staid in rather too long. . . . I shall be more careful another time."[11]

In some quarters, hydromania of this sort aroused suspicion for different reasons. The beach was a quasi-democratic space where people who led otherwise quite separate lives suddenly found themselves in close and awkward proximity; although customs and regulations varied from resort to resort, there was considerable scope for transgression of gender boundaries as men and women saw each other in various states of undress or even completely naked. The anonymous author of *Observations on Indecent Sea-Bathing* (1805) expresses his horror at the discovery that at several locations in Devon "the bathing-machines, though destitute of awnings, *are not separated*, and the sexes may be said to bathe promiscuously." He is also concerned that, at Brighton, men have taken to bathing with no clothes on in full view of the public – "a most unmanly insult to the Fair Sex."[12] The pamphlet takes the form of a letter to the *Sun*, the editor of which agrees that, although the right to bathe in the sea is common to all, "the right of bathing, whether in the sea or in a river, must be *decently* exercised" (11). First-hand evidence that, at Brighton, women have become habituated to the sight of men bathing in the nude – a spectacle that "neither drove them from the windows, nor prevented them from parading on the Cliff" (8) – leads to sombre reflections on the degeneration of the female character. It is also pointed out that in Napoleonic France a law has recently been passed prohibiting bathing in the Seine without bathing dresses: "Shall the legitimate, the free, the paternal Government of our gracious Sovereign," the author asks, "be destitute of *any* security, just and honourable in its nature, which is employed to give nobility to the ferocious and wide-spreading tyranny of the Gallican despot? At all events, shall the British Fair be familiarized to scenes, which are deemed too gross for the licentious females of profligate France?" (10–11). There was virtually no area of British life that was left untouched by the politics of the French Revolution and Napoleonic Wars; here, we see clearly how the democratic pastimes of sea-bathing and swimming got caught in the riptides of that tumultuous era.

Of course, what some saw as socially or morally retrograde, others found amusing or an object of libidinal investment. The caricaturist Thomas Rowlandson was particularly fond of scenes of river bathing, and gave full rein to his erotic fantasies in depicting scenes of naked men and women entwined on a river bank or swimming in the sea, as in his view of "Venus's Bathing" at Margate, overlooked by people on the cliffs behind (c.1800; Figure 5.2, see plate section). The fact that swimming naked was the norm – at least for men – until well into the nineteenth century conveys the impression that swimming was a perfectly natural thing to do, and that in this period it perhaps represented part of the Rousseauesque revolt against the chains of social convention. But the question of whether swimming was a natural or unnatural activity was a recurring issue in literature on the topic right from the start. In this regard, one remarkable aspect of the discourse on swimming in the Romantic era is the primitive and anachronistic state of instructional literature. Until the mid-1810s, there were really only two sources of advice in print for anyone interested in learning to swim. One was a short but much reprinted and quoted letter by Benjamin Franklin, which urges teaching swimming to children not only as a life-skill that will relieve them in certain situations from "painful apprehensions of danger," but also as a wholesome and enjoyable form of exercise.[13] As Scott Cleary has shown, Franklin's work on swimming married these educational objectives with his scientific interest in the properties of water and human biomechanics, as well as pride in his personal mastery of this demanding "art."[14] The other source was a book by Sir Everard Digby first published in Latin in 1587, then translated and abridged as *A Short Introduction for to Learne to Swimme* by Christopher Middleton in 1595. William Percey's *The Compleat Swimmer*, which appeared in 1658, was essentially a plagiarism of Digby's work, which was also translated into French by Melchisedech Thevenot in 1696 and back-translated into English as *The Art of Swimming* a few years later. These are all essentially the same work. At the end of the eighteenth century and in the early decades of the nineteenth such public advice on swimming as appeared in print was invariably still based on Digby and his translators, sometimes with the interpolation of Franklin's practical guidance. It is curious that, until the appearance of Frost's *Scientific Swimming* in 1816, textbooks on swimming were tirelessly rehearsing the views of an Elizabethan Cambridge scholar aiming to persuade a suspicious elite audience that swimming should form part of the education of a gentleman.

In Middleton's translation, Digby's work begins by establishing a firm philosophical foundation for his practical advice. He argues that, far from swimming being against nature, a human's ability to swim proves their superiority in the great chain of being: whereas a fish can do nothing but swim, a man can swim without being limited to that medium; moreover, he can perform all sorts of "fine feates" in the water that far exceed the capabilities of fish. "So fit is the constitution of mans body," he declares, "that who so dooth but with himselfe thoroughly consider of it, cannot but accord with mee in thys, that a man of all creatures under the circumference of heaven, naturally excelleth in swimming."[15] This line is repeated in the many translations and adaptations of Digby that appeared in the next two hundred years: thus William Percey maintains that man, "as he is the more Noble, and above all other Creatures; so indeed he excels them all in Swimming, nay Fishes themselves," while the anonymous compiler of an early nineteenth-century edition of *The Art of Swimming*, published around 1830 but still essentially a version of Digby, states that man not only swims as naturally as other creatures but "with more perfection and variety, both for pleasure and advantage."[16] Throughout the Georgian era, therefore, literature on swimming was bizarrely and anachronistically fighting a two hundred-year-old battle against the dominant Elizabethan view of swimming as unnatural, at the same time as visitors to seaside resorts around the English coast were confronting the spectacle of swimmers bathing *au naturel* and some of the leading writers of the Romantic period were sharing in this most democratic of pleasures and exploring its potential in prose and poetry. What is even more intriguing is that many of the "fine feates" described and illustrated by Digby and endlessly rehashed by his imitators – which include carrying two birds across a river, cutting one's toenails, and practising a ridiculous manoeuvre called "the leap of the goat" – have the reverse effect of making swimming appear contrived and *un*natural.

In the Romantic period, therefore, we have on the one hand a social practice of swimming accelerating in popularity and causing concern to the guardians of public morality, and on the other an instructional literature stuck in the discursive backwaters of the late sixteenth century. What of the poets and novelists who participated in the new water-lust? In the remainder of this essay I intend to focus mainly on poets and poetry, but it is worth noting in passing that swimming makes regular and often significant appearances in Romantic fiction. In Maria Edgeworth's *Belinda* (1801), for example, the male protagonist, Clarence Hervey, a man of fashion who likes to excel in everything and enjoy the praise of men and women alike, engages in a series of challenges with a male acquaintance. Having already beaten him at wine-tasting and a pedestrian race around Hyde Park, flush with success he accepts a further double-or-quits challenge to a swim in the Serpentine, despite the fact that his only knowledge of swimming is a distant memory of reading Benjamin Franklin's essay on the subject. Inevitably Hervey nearly drowns and is saved only by the ministrations of a passer-by who is evidently familiar with the principles of resuscitation disseminated by the recently-formed Humane Society; his embarrassing lack of aquatic prowess represents a considerable blow to his self-esteem and public image. Few of Clarence's fashionable acquaintance, it seems, can swim, and it is clear that the author of *Practical Education* views this as a deficiency. In Thomas Holcroft's

Jacobin novel, *The Adventures of Hugh Trevor* (1794), by contrast, the eponymous hero, a farmer's son, has been given a robust physical education by his father, having been "taught to sip ale, eat hung beef, ride like a hero, climb trees, run, jump, and swim."[17] The latter skill becomes instrumental in the plot when he rescues from drowning a man trapped in a carriage which has overturned into a river – this man later turning out to be his formerly estranged grandfather, with whom he is subsequently reconciled. Finally, in Mary Shelley's apocalyptic fiction, *The Last Man* (1826), we see, towards the end of the novel, the last three human survivors of a global epidemic of the plague trying to sail across the Adriatic Sea to Greece. A storm occurs and the boat is capsized. Two of the occupants, including Lord Adrian (a character loosely based on Percy Shelley, who famously could not swim), are drowned, but Lionel Verney, who has always relished a physical challenge and treats a hostile sea as an opportunity for egotistic self-assertion ("I loved to feel the waves wrap me and strive to overpower me; while I, lord of myself, moved this way or that, in spite of their angry buffetings"), makes his way to shore and becomes literally the "last man" of the book's title.[18] In all these examples swimming is used to differentiate between characters and is seen as both a valuable life-skill and the expression of a more vital masculinity.

It is in poetry, however, that swimming develops its most complex and intriguing range of meanings in the Romantic period. Charles Sprawson, in an appealing non-academic survey, has written of the fascination with water, or "love affair with 'moistness'" common to many nineteenth-century writers; he notes that the "passion for bathing really began with the Romantic generation," and that "'swim' was a word that particularly appealed to its poets."[19] This is a valuable insight, and I shall now look in more detail at three of the poets who feature in Sprawson's untidy narrative.

Coleridge, whose two most famous poems, "The Ancient Mariner" and "Kubla Khan," feature landscapes of ocean, ice, sacred river and mighty fountain, is a leading example of an enthusiastic Romantic swimmer. Coleridge learned to swim in the river Otter as a child in Devon; when removed to London as a charity schoolboy after his father's death he made illegal bathing expeditions to the New River to the north of the city; on a visit to the Yorkshire coast in 1801 to indulge his illicit love for Sara Hutchinson he bathed regularly and "frolicked in the Billows," ignoring medical advice to avoid the sea because he had "Faith in the Ocean";[20] during his stay in Malta he got up before sun-rise to swim in the sea; while in later years he visited the then fashionable resort of Ramsgate ten times between 1819 and 1833, generally for a month or two in the autumn, and delighted in the cold sea-bathing prescribed as part of his health regime.[21] At first he found his own private spot a mile along the shore where there was "a good roomy arched Cavern" in which he could deposit his clothes before enjoying "a glorious tumble in the waves." At other times he used the commercial bathing machines (as depicted in Benjamin West's 1788 painting of *The Bathing Place at Ramsgate* [Figure 5.3, see plate section]), relishing the rough seas in which he could dive into the biggest waves off the top step of the ladder and surrender to their power: "I watched each time from the top-step for a high Wave coming, and then with my utmost power of projection *shot* myself *off* into it, for all the world like a Congreve Rocket into a Whale."[22]

Coleridge's swimming is a curious mix of childlike physical exuberance and hypochondriacal obsession with the medicinal virtues of sea water. His poem "After Bathing in the Sea at Scarborough in Company with T. Hutchinson, August 1801," the product of his trip to the Northeast with Sara Hutchinson, conjures a range of emotions:

> God be with thee, gladsome Ocean!
> How gladly greet I thee once more —
> Ships and Waves and endless Motion
> And Life rejoicing on thy Shore.[23]

He expresses confidence in returning health and a proud superiority over the fashionable crowd who tremble at the prospect of cold water; but the "Thoughts sublime" (15), "Grieflike Transports" (18) and "Silent Adorations" (19) point to a more intense, characteristically Romantic response – all the more compelling in that these "revisited" feelings seem as potent as their originals. The "endless Motion" of the sea evidently inspired a surge of hope and creativity, as well as renewed spiritual conviction ("God is with me, God is in me" [23]), in one who was always prone to depression and self-pity – leaving aside the vicissitudes of his love life and the effects of his already established opium addiction.

Byron was, as my introductory remarks suggested, a very different kind of swimmer. He learned to swim as a boy in the river Don in Aberdeen; thereafter, wherever this well-travelled aristocrat found himself – Harrow, Cambridge, Brighton, Hastings, Portugal, Italy, Greece, Turkey – he swam, very often on a daily basis. His deformed right foot constituting much less of a handicap in the water than on land, swimming was clearly to some extent an escape from his acute consciousness of his disability. But the freedom and exhilaration Byron found in swimming went beyond such compensatory urges; for him, swimming was an opportunity for egocentric, competitive self-display. The sentiments he attributes to poetic alter-egos like Childe Harold and Manfred – the latter recalls how he loved

> to plunge
> Into the torrent, and to roll along
> On the swift whirl of the new breaking wave
> Of river-stream, or ocean, in their flow.
> In these my early strength exulted[24]

– closely resemble his own pride and joy in accomplishing the extraordinary swimming feats that fed the construction of his legend. The celebrated Hellespont swim, which he "plume[d]" himself on "more than I could possibly do on any kind of glory, political, poetical, or rhetorical,"[25] and to which he audaciously alluded in the shipwreck scene in canto 2 of *Don Juan*, was preceded in 1809 by his two-hour crossing of the river Tagus in Lisbon against a dangerous counter-current, and somewhat diminished by his epic race from the Venetian Lido to the far end of the Grand Canal in June 1818, when he was in the water for four and a quarter hours. While these were performances of Olympian bravado, Byron's swimming could also be a way of managing strong emotions, as when he broke away from the cremation of Shelley's body and swam out to his boat anchored in the bay – a round trip of around three miles.[26] In his final years prolonged swims seem to have been injurious to Byron's health, yet he persisted in the habit. More than with any other Romantic, swimming was an inseparable part of Byron's identity, as his Italian neighbours recognized in erecting a plinth referring to him as a "noted English swimmer and poet," seemingly according equal importance to both occupations or activities.[27]

When considering the obvious "hydromania" of individuals such as Coleridge and Byron, it is tempting to offer broad speculative explanations for the surge in popularity of swimming among the Romantic generation. Like walking, which boomed as a middle-class recreational practice during this period, swimming was an accessible, everyday activity that appealed to those inspired by the democratic and egalitarian ideals of the French Revolution. Although it possessed a nascent infrastructure (in the form of bathing facilities at coastal resorts) and eventually became heavily regulated and codified, for many young Romantics swimming was an expression of personal freedom and spontaneity and as such their passion for open water, as I shall argue below, finds a direct descendant in today's wild swimming movement. Unlike walking, the state of relative or total undress associated with swimming had the potential to give group participation in the latter a mildly subversive erotic subtext – a dimension of which caricaturists and printmakers of the period were keenly aware. For those enthused by the Romantic discovery of natural landscape and the flourishing trend in landscape art, swimming provided a literally immersive experience of nature and the thrilling possibility of blurring the boundary between spectator and spectacle on which the aesthetics of picturesque beauty and the sublime both depended in different ways. For some men, such as Byron, swimming allowed for the performance of a version of masculinity based on physical prowess and competitive athleticism – a counterweight to the effeminizing tendencies of the contemporary cult of sensibility.

Whatever the specific character of the passion in individual cases, there is no doubt that the rapid rise of recreational swimming led to its greater prominence in a range of literal and figurative contexts in Romantic literature. I have already glanced at some fictional instances; I would now like to look at some striking poetic examples. Coleridge's poem, "This Lime-Tree Bower My Prison," in which the incapacitated speaker goes on a mental walk with his friends through scenery with which he is evidently well acquainted, reaches its rhetorical climax as they watch the sunset from an elevated viewpoint:

> So my Friend
> Struck with deep joy may stand, as I have stood,

Silent with swimming sense; yea, gazing round
On the wide landscape, gaze till all doth seem
Less gross than bodily; and of such hues
As veil the Almighty Spirit, when he makes
Spirits perceive his presence.[28]

Here, "swimming sense" denotes a temporary disorientation (a sensation aurally reinforced by the overwhelming sibilance of these lines, in which sound threatens to occlude signification), a suspension of ordinary sense perception, and the threshold of a religious experience. Familiar things lose their material shape and become more spiritual; even the spectator loses his physical identity and becomes a spirit in the presence of a higher Spirit. It is fascinating that swimming is the chosen metaphor for this apparent merging of self and world, this unsustainable moment of revelation.

There is a comparable moment in another of Coleridge's conversation poems, "Frost at Midnight." Sitting up at night in his cottage, listening to the quiet breathing of his baby son, the speaker observes a sooty film fluttering on the grate and is mentally transported back to his unhappy schooldays, when he often observed a similar phenomenon and eagerly anticipated the arrival of the stranger – a family member, perhaps – that a popular superstition indicated was thereby imminent:

So gazed I, till the soothing things, I dreamt,
Lulled me to sleep, and sleep prolonged my dreams!
And so I brooded all the following morn,
Awed by the stern preceptor's face, mine eye
Fixed with mock study on my swimming book:
Save if the door half opened, and I snatched
A hasty glance, and still my heart leaped up,
For still I hoped to see the *stranger's* face,
Townsman, or aunt, or sister more beloved,
My play-mate when we both were clothed alike![29]

Here, it is the "swimming book" that most vividly conveys the speaker's inattention to his familiar surroundings and his withdrawal into an inner world of reverie and brooding anticipation. As with the previous example, swimming is a potent metaphor for the erasure of everyday reality and the entry into a trance-like state that promises – even if it does not here deliver – a fulfilling, transformative experience.

The poetry of Byron offers a crowded mosaic of literal and figurative references to swimming. Here I shall glance solely at *Don Juan*, in which the author's aquatic imagination is constantly associating the act of swimming with intense emotions and liminal situations. I have already mentioned the famous shipwreck scene in canto 2, in which Byron explicitly aligns his own capabilities in the water with those of his hero:

A better swimmer you could scarce see ever,
He could, perhaps, have pass'd the Hellespont,
As once (a feat on which ourselves we prided)
Leander, Mr Ekenhead, and I did.
(*Don Juan*, 2.837–40)

Byron's insistence that his Hellespont crossing proved the practicability of Leander's mythical exploits – that "a young Greek of the heroic times, in love, and with his limbs in full vigour, might have *succeeded* in such an attempt," as he declared in a letter published in the *London Magazine*[30] – suggests the close connection that he posited between athletic and erotic prowess. This link is strengthened in the ensuing description of Juan's relationship with Haidée, a Greek pirate's daughter. The two meet when Juan, the only one of four emaciated survivors from a capsized lifeboat to have made his way to land (one of his companions is eaten by a shark, the other two cannot swim), regains consciousness after having collapsed on the beach. He now sees his new love-interest for the first time: "slowly by his swimming eyes was seen/A lovely female face of seventeen" (2.895-96). While the "swimming eyes" here are at least partly a reference to Juan's state of dizziness, later in the poem we find swimming used in a fully figurative sense:

Juan and Haidée gazed upon each other
With swimming looks of speechless tenderness,
Which mix'd all feelings, friend, child, lover, brother,
All that the best can mingle and express
When two pure hearts are pour'd in one another,
And love too much, and yet can not love less.
(4.201–6)

Here, swimming is linked to confusion and disorder – it is a love that "mix'd all feelings, friend, child, lover, brother" – and is marked as excessive and transgressive. In the image of "two pure hearts . . . pour'd in one another," water is the solvent of propriety as well as of personal identity. And whereas Coleridge's "swimming sense" was the gateway to a religious epiphany, Byron's "swimming looks" betoken a different form of transcendence achieved via sexual union: "Love was born *with* them, *in* them, so intense,/It was their very spirit—not a sense" (4.215–6).

Byron applies the metaphor of swimming to other kinds of disorientation or self-estrangement. There is an interesting occurrence in the opening stanzas of canto 9, where the narrator toys with Hamlet's famous question, "To be or not to be," and alludes to a long history of philosophical scepticism stretching back to ancient Greece:

It is a pleasant voyage perhaps to float,
  Like Pyrrho, on a sea of speculation;

But what if carrying sail capsize the boat?
   Your wise men don't know much of navigation;
And swimming long in the abyss of thought
   Is apt to tire: a calm and shallow station
Well nigh the shore, where one stoops down and gathers
   Some pretty shell, is best for moderate bathers.

(9.137–44)

There is a playful echo in these lines of the episode in canto 2, but it is a form of mental shipwreck that Byron here contemplates and he is not so confident of his ability to get to shore in these circumstances. It is a nice irony that Byron, while using open-water swimming as a metaphor for the pointless, self-disconcerting character of philosophical abstraction, refers to himself – albeit through the persona of the down-to-earth, practically-minded narrator – as a "moderate bather."

That modest appellation would probably be more appropriately applied to John Keats, although hard evidence of the latter's water-based activity is difficult to come by. He seems to have swum in a pond in the grounds of his school in Enfield and availed himself of the nearby New River at various times for the same purpose. On his northern walking tour in 1818 he definitely bathed in a "quite pat and fresh" Loch Fyne.[31] Exactly how good a swimmer Keats was is impossible to determine, but perhaps more important is the fact that habitually, for Keats, "water created a glorious torrent of poetry."[32] That torrent contains numerous fascinating swimming images and metaphors. In "I stood tip-toe" there is a beautiful picture of the moon "lifting her silver rim/Above a cloud, and with a gradual swim/Coming into the blue with all her light."[33] In the climactic scene of *Lamia* where the female protagonist is exposed as a serpent, the narrator pleads for her fiancé Lycius to be spared his disillusionment and pain via some kind of drunken oblivion: "let us strip for him/The thyrsus, that his watching eyes may swim/into forgetfulness" (2.225–7). However, in the space remaining I have room to comment only on the most famous introduction of swimming into Keats's poetry.

In his early sonnet, "On First Looking into Chapman's Homer," Keats begins by using an overarching metaphor of travel to describe his reading of poetry, then switches to a different image to convey his intense delight at perusing for the first time an Elizabethan English translation of Homer's epic poetry:

Oft of one wide expanse had I been told
   That deep-brow'd Homer ruled as his demesne;
   Yet did I never breathe its pure serene
Till I heard Chapman speak out loud and bold:
Then felt I like some watcher of the skies
   When a new planet swims into his ken.

(5–10)

The thrill of the astronomer discovering a new planet reworks the language of exploration used in the first eight lines of the poem. But "swims" introduces a different element: it is hard to imagine a planet swimming, except in so far as it might appear to be floating in some invisible aqueous element, but easier to imagine the "swimming sense" or consciousness of the astronomer, unable to believe what he is seeing. Swimming again signifies a moment of extraordinary insight when old certainties dissolve and a previously knowable universe becomes more fluid and indeterminate. It is remarkable that Keats consistently reaches for a swimming metaphor when he wants to describe unusual states of mind – whether in a positive way, as here with the representation of a classic experience of the sublime, or in a more negative context, as when he describes his failure to make progress with *Endymion* owing to an "overwrought" brain. The "swimming in [his] head]" is compared to "the effects of a Mental Debauch", producing nothing but "anxiety to go on without the Power to do so."[34] From early modern times onwards, as we have seen, swimming has connoted a blurring of boundaries; for Keats it sometimes gestures towards an exciting and productive suspension of normality, while here it signifies more of an intellectual and moral limbo.

In conclusion, the passion for swimming – the hydromania – that developed in the Romantic era is a subject that has received informed coverage in books aimed at the general reader, but has attracted little or no scholarly attention. In this essay I have touched on several aspects of a complex phenomenon. In my remarks on creative writers I have shown how swimming was a liberating, confronting and sometimes transgressive enthusiasm. Water is a transforming medium, and swimming was increasingly valued as a transformative experience; its literary uses were heavily coloured by connotations of change and renewal, of dissolving boundaries and fluid identities. In describing his marathon swim in Venice, Byron declares that he is "almost amphibious" and adds: "If I believed in the transmigration of your Hindoos, I should think I had been a *Merman* in some former state of existence, or was going to be turned into one in the next."[35] Modern proponents of the neoromantic wild swimming movement pioneered by Kate Rew, Daniel Start, Rob Fryer and others talk in not dissimilar terms. Rew, for instance, reflects that there "comes a point if you do front crawl continuously for an hour you kind of lose the sense of your joints. You become quite fish like. You feel like muscle just flexing your way through the water." Daniel Start talks of the thrill of swimming in white water courses like the river Dart in Devon, where "You have to very much give yourself over to the forces and move like an eel or an otter with the way the water is flowing and with just a flick of your flippers or your hands you move in and out. Become part of the flow."[36]

The godfather of modern wild swimming is, however, the late writer and filmmaker Roger Deakin, whose influential *Waterlog* (1999) was itself inspired by the well-known short story, "The Swimmer," by the American writer John Cheever. Cheever's protagonist Ned Merrill decides one day to take the eight-mile journey home across Long Island by traversing every swimming pool along the way, following what his cartographic imagination sees as a "quasi-subterranean stream that curved across the county." Merrill has "an inexplicable contempt for men who did not hurl themselves into pools," and regards the "embrace" of the light green water as "the resumption of a natural condition."[37] On a much larger scale, Deakin's journey takes him the length and breadth of Britain, immersing him in seas, rivers, lakes and ponds in a rambling, watery picaresque adventure. He too writes of swimming as a transformative experience, valuing the dissolution of his terrestrial identity:

> Swimming is a rite of passage, a crossing of boundaries: the line of the shore, the bank of the river, the edge of the pool, the surface itself. When you enter the water, something like metamorphosis happens. Leaving behind the land, you go through the looking-glass surface and enter a new world, in which survival, not ambition or desire, is the dominant aim. . . . You see and experience things when you're swimming in a way that is completely different from any other. You are *in* nature, part and parcel of it, in a far more complete and intense way than on dry land, and your sense of the present is overwhelming. In wild water you are on equal terms with the animal world around you: in every sense, on the same level.[38]

Here, as with the reflections of Rew and Start, we are taken all the way back to the world of the Elizabethans and to the underlying anxieties of Sir Everard Digby's pioneering instruction manual: is it natural for a man (or woman) to swim? Do we belong in the water? Is a human being better than a fish?

But of course the discourse has shifted in a fundamental way. Swimming, we still feel, troubles the boundaries between man and nature, human and animal – troubles our sense of self; but we now embrace those uncertainties and instabilities. In the subaquatic world which is the main focus of this collection of essays the kind of voluntary, barely comprehended metamorphosis that Deakin describes takes on still greater potency, as do the sensuous and phenomenological rewards of which it is productive. But before this much later development could take place, men and women had to learn first to appreciate the emancipatory and transgressive potential of swimming, of crossing the line of the shore and breaking the surface of the water. In shaping that distinctively modern sensibility the Romantic generation, as in so many areas, took the first decisive strokes.

Notes

This chapter previously appeared as an article in *Romanticism*, and is reprinted with the permission of Edinburgh University Press. Robin Jarvis, "Hydromania: Perspectives on Romantic Swimming." *Romanticism* 21.3 (2015): 250–64 © Edinburgh University Press.

1 The other two prints are reproduced in C. Wanklyn, *Lyme Regis: A Retrospect* (London: Hatchards, 1927).
2 Becky Horsbrugh, "The Hellespont Swim: Following in Byron's Wake," *The Guardian*, May 6, 2010, http://www.theguardian.com/lifeandstyle/2010/may/06/hellespont-swim-byron.

3 Southey to Thomas Phillipps Lamb, April 3, 1793, in *The Collected Letters of Robert Southey Part One: 1791–1797*, eds. Lynda Pratt, Tim Fulford, and Ian Packer, Romantic Circles Electronic Edition, Appendix 1, Letter A, http://www.rc.umd.edu/editions/southey_letters/letterEEd.26.appendix1.html.
4 Coleridge to Thomas Allsopp, December 3, 1821, in *The Collected Letters of Samuel Taylor Coleridge*, ed. E. L. Griggs, 6 vols. (Oxford: Oxford University Press, 1956–71; henceforth *CLSTC*), 5:188.
5 Keith Thomas, *Man and the Natural World: Changing Attitudes in England 1500–1800* (London: Penguin, 1983), 38–9.
6 Michael West, "Spenser, Everard Digby, and the Renaissance Art of Swimming," *Renaissance Quarterly* 26, no. 1 (1983): 14.
7 I have explored the Romantic cult of the walking tour in my *Romantic Writing and Pedestrian Travel* (Basingstoke: Macmillan, 1997). On mountaineering, see Simon Bainbridge, "Romantic Writers and Mountaineering," *Romanticism* 18, no. 1 (2012): 1–15.
8 Thomas Reid, *Directions for Warm and Cold Sea-Bathing; with Observations on their Application and Effects in Different Diseases*, 2nd edn. (London: Cadell and Davies, 1798), 4.
9 Susie Parr, *The Story of Swimming: A Social History of Bathing in Britain* (Stockport: Dewi Lewis Media, 2011), 44–7.
10 Jane Austen, *Sanditon*, ed. Margaret Drabble (London: Penguin, 1974), 163.
11 Austen to Cassandra Austen, September 14, 1804, in *Jane Austen's Letters*, ed. Deirdre Le Faye, 4th edn. (Oxford: Oxford University Press, 2011), 99.
12 *Observations on Indecent Sea-Bathing, as Practised at Different Watering-Places on the Coasts of this Kingdom* (London: J. Hatchard, 1805), 5, 8.
13 Benjamin Franklin, *Experiments and Observations on Electricity, Made at Philadelphia in America* (London, 1769), 468.
14 Scott Cleary, "The Ethos Aquatic: Benjamin Franklin and the Art of Swimming," *Early American Literature* 46, no. 1 (2011): 51–67.
15 Christopher Middleton, *A Short introduction for to Learne to Swimme* (London, 1595), n.p.
16 William Percey, *The Compleat Swimmer: Or, the Art of Swimming* (London, 1658), Preface, n.p.; *The Art of Swimming Made Safe and Easy* (London, c.1830), 5.
17 Thomas Holcroft, *The Adventures of Hugh Trevor*, ed. Seamus Deane (Oxford: Oxford University Press, 1973), 17.
18 Mary Shelley, *The Last Man*, ed. Morton D. Paley (Oxford: Oxford University Press, 1994), 442–3.
19 Charles Sprawson, *Haunts of the Black Masseur: The Swimmer as Hero* (London: Vintage, 1992), 32, 134.
20 Coleridge to Robert Southey, 11 August 1801, *CLSTC*, 2:751.
21 See Allan Clayson, "'Wish You were Here': The Significance of Coleridge's Holidays at Ramsgate," *The Coleridge Bulletin*, n.s., 16 (2000): 15–23.
22 Coleridge to James Gillman, c.19 August 1819, *CLSTC*, 4:946; Coleridge to James Gillman, 31 October 1821, *CLSTC*, 5:185.
23 *Poetical Works* (henceforth *PW*), ed. J. C. C. Mays, in *The Collected Works of Samuel Taylor Coleridge* 16, 3 vols. in 6 parts (Princeton: Princeton University Press, 2001), vol. 1, part 2, 658, lines 1–4. Subsequent line references will be incorporated parenthetically in the text.
24 *Manfred*, 2.2.65–9. All quotations from Byron's poetry are from *Byron*, ed. Jerome J. McGann, Oxford Authors (Oxford: Oxford University Press, 1986). Subsequent references will be incorporated parenthetically in the text.
25 Byron to Francis Hodgson, July 4, 1810, *Byron's Letters and Journals*, ed. Leslie A. Marchand, 13 vols. (London: John Murray, 1973–94), 1:253.

26 See Fiona MacCarthy, *Byron: Life and Legend* (London: John Murray, 2002), 430–1.
27 Sprawson, *Haunts*, 102.
28 *PW*, vol. 1, part 1, 352–53, lines 37–44.
29 *PW*, vol. 1, part 1, 455, lines 34–43.
30 "Swimming across the Hellespont: Letter from the Right Honourable Lord Byron to Mr Murray," *The London Magazine* 3 (April 1821): 364.
31 Keats to Tom Keats, July 17–21, 1818, in *Letters of John Keats*, ed. Robert Gittings (Oxford: Oxford University Press, 1970), 130.
32 Nicholas Roe, *John Keats* (New Haven: Yale University Press, 2012), 239.
33 *The Poems of John Keats*, ed. Jack Stillinger (London: Heinemann, 1978), 83, lines. 113–15. All references to Keats's poetry are to this edition; subsequent line references will be incorporated in the main text.
34 Keats to Taylor and Hessey, May 16, 1817, *Letters*, 16.
35 Thomas Medwin, *Conversations of Lord Byron: Noted during a Residence with his Lordship at Pisa, in the Years 1821 and 1822* (London: H. Colburn, 1824), 137, 139.
36 Quoted in Parr, *Story of Swimming*, 172.
37 John Cheever, "The Swimmer," in *Collected Stories and Other Writings* (New York: Library of America, 2009), 727.
38 Roger Deakin, *Waterlog: A Swimmer's Journey through Britain* (London: Chatto & Windus, 1999), 3–4.

6 The great melancholy mother

Michelet's evolutionary ocean in *The Sea*

Natalie Deam

A significant cultural shift occurred in the mid-nineteenth century as popular fascination with the ocean moved away from the sublime surface of the waves towards the strange biologies lurking in the submarine. This chapter will examine the shift from the Romantic ocean to what I will call the evolutionary ocean by illustrating the ways that popular scientific texts like Jules Michelet's 1861 natural history *The Sea* (*La Mer*) changed the representation of the relationship between man and the marine world. This shift depended in part on developments in technologies such as the microscope, diving bell, and aquarium that made marine flora and fauna increasingly accessible to scientists. Thanks also to aquariums at the World's Fairs, natural history exhibits of marine specimens, and other public aquatic spectacles, popular fascination with marine biology helped to pull increasing attention toward the teeming ocean depths. Recent cultural histories of the ocean such as Natascha Adamowsky's *The Mysterious Science of the Sea*, Alain Corbin's *The Lure of the Sea*, and Judith Hamera's *Parlor Ponds* reveal how modern technologies spectacularized the biological curiosities of the nineteenth-century ocean.[1] Scholars have focused less, however, on the imaginative marine biology found in popular scientific texts like Michelet's *The Sea*, Alfred Frédol's *The World of the Sea* (*Le Monde de la mer*) (1865), Frédéric Zürcher and Elie Philippe Margollé's *The Underwater World* (*Le Monde sous-marin*) (1868), Arthur Mangin's *The Mysteries of the Sea* (*Les Mystères de la mer*) (1872), or Henri Filhol's *Life at the Bottom of the Sea* (*La Vie au fond des mers*) (1885). As Michelet's Romantic natural history explicitly treats evolutionary marine biology, a close reading of this text offers unique insight into this cultural shift and the many anxieties and fantasies it produced. Using Michelet's *The Sea*, I will show how nineteenth-century texts complicated Romanticism's easy identification with the sea by challenging readers to recognize man's own evolutionary connections to the submarine. As part of the enormous impact that evolutionary theory had on refiguring nineteenth-century understandings of nature, the emerging discourse of the evolutionary ocean shows how the strangeness of marine biology challenged traditional conceptions of the sea, humanity, and our relationship to the natural world.

In his monumental history of the French oceans, *The Lure of the Sea*, Alain Corbin claims that the Romantics were the first to create a coherent discourse of the seaside through routine visits to the coasts, where they explored connections between the sublime seascape and the human spirit.[2] Within the idealized Romantic landscapes Corbin describes, there is little room for the human spectator to reflect on other living creatures, however. The Romantics remained engaged with the fantasy of the deserted surface and empty abyss rather than with the creatures that resided beneath the waves. Indeed, the Romantic ocean served as both a mirror of the surrounding landscape and a reflection of the poet's own aesthetic and emotional complexity. Take Victor Hugo's seaside reflections in "Beside the Sea" ("Le Bord de la mer"): "And so! cloud, azures, space, ether, abyss/This fluid ocean, these regions sublime/Full of fires, of glimmers, of beams/Where the soul takes man, where we both flee."[3] Here Hugo effortlessly blends sea, sky, heavens, and depths into a limitless sublime that only those endowed with the poet's ability to embrace the infinite could recognize. This Romantic anthropomorphism of the marine sublime is especially visible in Baudelaire's "Man and the Sea" ("L'Homme et la mer"): "Free man, you will always cherish the sea!/The sea is your mirror; you contemplate your soul/In the infinite unrolling of its waves/Your spirit is no less bitter an abyss."[4] Baudelaire's emotional connection to the empty abyss reveals the depths of his soul; however it is an idealized and uninhabited abyss more concerned with melancholic grandeur than biology. Moreover, Baudelaire's poem reveals that as much as Romantic marine discourse evokes the magnanimity of nature, it also demonstrates the poet's own sentimental and aesthetic prowess. Rather than losing the poet within the sublime marine depths, Baudelaire's analogies flatter the poet by expanding his soaring ideals and tortured soul to encompass oceanic infinitudes of space and time.

Romantic representations of the marine sublime almost always foreground humanity's ability to understand and represent the ocean at such a scale, undoubtedly compensating for our inability to control, conquer, or inhabit the sea. Yet a careful look at the ocean in mid-century popular science reveals that a more complicated engagement with the marine emerged in the wake of evolutionary theory: one that recognized the sea as an inhabited, albeit emphatically non-human space, with which modern man had difficulty identifying. This shift depends first on imaginatively populating the ocean depths with animal life. Corbin explains that historically, the ocean had been imagined as a timeless desert: "Towards the end of the eighteenth century ... the ocean would call forth the image of a 'vast expanse' that was indifferent to human time, like the desert; a place of sublime *vacuity* whose imagined depth was modeled on the very perpendicular sides of the mountains that often bordered it."[5] In *The Sea* Michelet cites Darwin to revivify the empty ocean, transforming the depths into an hallucinatory display of life:

> 'The fields and forests of our dry land,' says Darwin, 'appear sterile and empty, if we compare them with those of the sea.' And, in fact, all who traverse the marvellous transparent Indian seas are thrilled, stirred, startled, by the phantasmagoria that flashes up from their far clear depths.[6]

Much like Darwin, Michelet includes extravagant lists and baroque descriptions of marine flora and fauna and often gestures to his inability to represent adequately the sea's incredible biodiversity. In order to make sense of the sea's strange forms, Michelet attempts to connect marine biology to more recognizable forms of terrestrial life. Yet his imaginative attempts to explain how complex lifeforms might have evolved from life's primordial marine origins demonstrate how in the era of Darwin, the reflections that modern man found peering back from the ocean depths were less flattering and far stranger than those of the Romantics' mirrored sea.

Jules Michelet, a Romantic historian famous for his national histories, was one of the first to attempt a cohesive evolutionary history of the ocean. Although his natural histories have perhaps been overlooked by scholars due to their heavy rhetoric and extravagant metaphors, these texts offer unique insight into the complicated reception of evolutionary theory and its deep connections to emerging marine biology. Written for popular audiences rather than academics, *The Sea* synthesizes recent theories of evolution, spontaneous generation, speciation, and extinction while highlighting the more spectacular specimens of marine biology including corals, sharks, and giant squid. Michelet indulges a metamorphic aesthetic of marine life in which the sea constantly recycles the material traces of its evolutionary history through the microscopic and monstrous organisms that compose the submarine world. Given the fact that Michelet wrote *The Sea* while he worked on his chef d'oeuvre *The History of France* (*Histoire de la France*), his fantastic exaggeration of the sea's biology may seem surprising. Yet his natural histories were far more lucrative than the national history for which he is now remembered. Indeed, *The Sea* outstripped *The History of France* in sales, quickly selling out of its first 24,000 copies; its immediate English translation also found similar success.[7] Although literary scholars often ignore the natural history genre, *The Sea*'s success evidences massive popular interest in marine biology and the importance of the genre in founding the discourse of the evolutionary ocean.

The practice of reading history in nature did not necessarily come easily to Michelet, however. In his provocative analysis of the parallels between Michelet's national and natural histories, Lionel Gossman focuses on Michelet's Romantic attempts to conform the human and natural world into a coherent narrative of progress:

> History, for Michelet, is thus nothing less than the never-ending struggle against the ancien régime of nature, the process by which nature, woman, and the past, in their confusing multitude of unstable, constantly varying forms, are progressively transformed from capricious mistresses of human destiny into trained assistants in the creation of a specifically human order.[8]

Although correct in describing Michelet's progressive view of history and nature, Gossman's reading ignores Michelet's frequent complications of this progressive view in *The Sea*. Moreover, Gossman does not account for how frequently Michelet casts doubt on his ability to understand the natural world. Despite diverse citations of evolutionary theorists and marine biologists including Charles Darwin, Matthew Maury, Bory de Saint-Vincent, and René de Réaumur, Michelet often questions his ability to correctly perceive the ocean. Consider his hallucinatory response to a coral reef:

> This glorious sight inspires, yet agitates us; it is a dream, a vertigo; that Fay of the shifting mirage, the Sea, adding to these colors her own prismatic tints, fading, reappearing, now here, now gone, a capricious and fitful inconstancy, a hesitation, a doubt. Have we really seen it, this lovely scene? No, it was not so. Was it an entity, or a delusion? Yes, yes, it must be real, there are certainly very real beings there, for I see whole hosts of them lodged there and sporting there. ... Strange fish, vast and curious monsters of the deep, move hither and thither in their many colored vesture of purple and gold, and deep azure and delicate pink.
>
> (143)

Rather than the Romantic sublime of desolate seascapes, Michelet evinces a new type of biological sublime provoked by the sheer scale of marine biodiversity and direct confrontations with the confusing, alien forms of the sea. As he discusses increasingly complex biologies, his evocation of this biological sublime intensifies, along with his doubts about reconciling a progressive evolution with the stranger lifeforms of the sea. Unable to comprehend the reef's dazzling flora and fauna – especially those organisms whose forms constantly change – Michelet hesitates to apply such a reductive history that could not account for the enduring diversity of marine life. A close reading reveals that Michelet's history reflects the same widespread anxieties that evolutionary theory produced around the world as it asked man to accept monstrous biologies not only as natural but also as products of the same evolutionary forces that led to the human. Although we often think of man's relationship to apes as the primary cause for evolutionary concerns, *The Sea* reveals that modernity also struggled to reckon with humanity's evolutionary origins in the slippery lifeforms of the sea.

Michelet's penchant for bizarre biology explodes in the second book of *The Sea*, *The Genesis of the Sea* (*La Genèse de la mer*), which narrates a progressive evolutionary history from the origins of life in warm volcanic shallows up through marine mammals to the unexpected appearance of man. To structure the sequence of chapters, each of which focuses on a different genus or species, Michelet uses the traditional – albeit outdated – model of the chain of being popularized by eighteenth-century natural historians. *The Genesis of the Sea* therefore reads as a history of the evolution of anthropomorphic physiology, wherein the comparative study of marine organisms reveals how human features, including eyes and hands as well as complex social structures, emerged over time from the marine depths.

Obvious tensions arise between the chain of being's static representation of nature and the fluctuating nature of an evolutionary worldview, which are further accentuated by Michelet's Romantic attempts to read individualism, liberty, and will within the natural world. Moreover, his mixed references to the conflicting evolutionary theories that circulated in nineteenth-century France further complicate a precise explanation of Michelet's understanding of evolution. In short, Michelet's progressive history borrows from Geoffroy Saint-Hilaire's deterministic and transcendental faith in the transmutation of a species across time, wherein nature slowly works out the potential of a species. Yet Michelet also exaggerates the voluntarism of Lamarckian evolution, as Edward Kaplan poignantly explains:

> The materialist explanation cannot satisfy the intransigent spiritualism of Michelet, which discerns a certain voluntarism across animal evolution. In order to show a conscious will in his work he makes use of Lamarck's famous principle: that an animal's organs can undergo modification in order to accomplish some need and that through habitual use these changes can become second nature. Add to that the notion of desire (or of interior feeling). . . . Michelet goes even beyond this. He postulates that the will is at the origin of all essential physical transformations.[9]

Although Michelet appreciates the deep scale of marine history, he depicts evolution occurring on a relatively rapid time scale, driven by individual need and an unconscious desire for humanoid characteristics. Yet the roles that Nature, the sea, and the individual play in this evolution remain ambiguous – especially because of his penchant for dramatic language.

Consider Michelet's enigmatic statement: "Each polyp is not resigned to remain a polyp" (167). Michelet's rhetoric underscores the metamorphic quality of all sea life; it also admits an ambiguous amount of self-determination for each organism. It is difficult to tell whether to read this as the demonstration of individual agency or as some complex co-evolutionary interaction between the organism and its environment. In short, it is unclear whether nature willfully or unconsciously directs the evolution of its organisms, or whether individuals determine their own forms and relationships. Moreover, when one considers the power of the polyp to reform underwater landscapes from isolated colonies to treacherous reefs to monumental atolls, Michelet's uncanny destabilization of the single polyp threatens to undermine any sense of enduring marine form as well as any explanation of its development.

It is no coincidence that the sea's liquid environment provoked Michelet to imagine an extremely flexible evolutionary history of marine life. His descriptions of metamorphic marine life forms lend themselves to slippery discourses of evolutionary transformation across *The Genesis of the Sea*. For example, in his chapter on marine mammals Michelet claims that the whale longs to hold her offspring but lacks the means to cradle her child. To better care for its young, the whale therefore transforms into what becomes the seal:

> Weight, so fatal to the whale, is much more so to this specimen. Let us thus reduce the size even more, cut back the rotundity, soften the spine— above all let us get rid of this tail, or rather dissolve the fork into two fleshy appendages that will be much more useful. . . . Everything liberates itself through constant effort. Nature strives towards caressing the child, approaching it and holding it. Ligaments cede, extend, let loose the forearm, and from this arm spreads a webbed polyp. — It's the hand.[10]

Michelet's series of active verbs accelerate the physiological changes that a more rigorous scientist would explain evolve over centuries. Although these nimble physiological transitions and verbs of transformation read as a much more Lamarckian than Darwinian account of evolution, as Michelet shifts from the first person imperative to an impersonal indicative he obscures the agency of evolutionary nature. It is unfeasible to discern Michelet's precise evolutionary inspirations from such figurative rhetoric; however, the discursive intersections of the emerging fields of evolutionary and marine biology and the material influence of the submarine on his natural history are clear. Michelet's dynamic, liquid representations of adaptation and speciation encourage readers to imagine a fantastic metamorphic marine in which any organism might rapidly perform the physiological transformations that structure his evolutionary history.

The Sea's imaginative descriptions of the sea push evolutionary theory to its limits. Further complicating a progressive reading of evolutionary marine biology, Michelet's history of the adaptation of anthropomorphic features elicits connections between the human and the non-human that become increasingly unnatural. Michelet's constant anthropomorphic comparisons, which ask readers to imagine the sea slug, coral polyp, or octopus as mankind's evolutionary ancestor, therefore provoked understandable anxieties. Confronting each new creature in the sea, Michelet seems to ask not "how does this resemble man?" as Hugo and Baudelaire ruminated before the seaside, but rather "how could man have evolved from *this*?" Yet despite his own insistence on the increasingly anthropomorphic forms inhabiting the sea, Michelet agonizes over identifying a precise genealogical relationship. The closer he comes to identifying the exact transition from non-human to human life, the more elusive he becomes:

> The Sea, no doubt, commenced everything. But it is not from the highest marine animals that has proceeded the long parallel series of terrestrials that is culminated and crowned in Man. They were already too fixed, too special, to form the first rude sketch of a nature so different. They had carried far, almost exhausted, the fecundity of their species. In that case the elders perish; and it is very low down in the obscure juniors of some parent class that the new series commences that is to ascend so much higher.
>
> (246)

This fleeting concession that the ocean is too primitive an environment to permit the evolution of man arrives late in *The Genesis of the Sea* – especially considering that the anthropomorphic structure of Michelet's entire history has insisted on this possibility. This structural conflict as well as the ambiguous rhetoric, theoretical leaps, and fantastic extremes to which Michelet resorts demonstrate the struggles the nineteenth century faced in identifying the human within the evolutionary ocean.

As fascination with marine biology and evolutionary theory grew, popular scientific texts such as *The Sea* experimented with how compatible the two emerging disciplines truly were. Michelet's natural history shows how this combination provoked anxieties over humanity's marine origins and an uncanny imagination of the metamorphic evolutionary marine. Natascha Adamowsky's extensive study of nineteenth-century representation of the marine underscores the ways that evolutionary biology stimulated this imaginative response: "The more knowledge of evolutionary biology increased, for example, the more one marveled at the wondrous works of nature: the overflowing creativity that had the anarchic power to produce unimaginable multiplicity."[11] Michelet's sense of wonder is readily apparent in his fanciful descriptions of sea life; however wonder cannot account for his uneasy attitude towards the extreme diversity and plasticity of marine life. His marvel often sours to concerns about the grotesque lifeforms we must accept as natural products of evolution, which cause Michelet to question our ability to understand scientifically the natural world at all. Rather than dismissing his uncertainties on sensory or technological shortcomings, however, Michelet indulges these doubts. His concerns therefore resonate with modernity's deep re-evaluation of mankind's historical and ecological situation within the natural world. Futilely probing the ocean depths for proof of humanity's origins also allows Michelet to realize the precariousness of the sea's biodiversity and the disastrous impacts of the whaling industry on marine life. Across *The Sea* he becomes increasingly concerned with how humanity might help maintain the ecological welfare of the sea.

Michelet's ecological concerns for the conservation of marine life depend not only on populating its depths but also understanding the ocean through organic metaphors. For instance, reflecting on the sublime currents that circle the globe pushes Michelet to consider the animality of the sea itself: "Look at a great clock, or a steam engine which imitates almost exactly the movement of the vital forces. Is that [the ocean] a freak of nature? Should we not far rather imagine that in these masses there is a mixture of animality?" (61) Here, animality introduces a sense of vital energy into the ocean that works against strict scientific understanding. Michelet's remarkable representations of the ocean as a giant organism demonstrate the power of marine biology to refigure the nineteenth century's understanding of the natural world. Asking readers to understand the ocean as an animal rejects traditional mechanistic visions of the natural world for a more fantastic imagination of nature, one which animates even water itself.

According to Michelet, sea water, which in a pun on the French word "*mer*" (sea) and "*mère*" (mother) he calls the "milk of the sea," is made up of the semen and eggs of asexual organisms, the blood and breastmilk of marine mammals, plumes of unclassifiable microscopic organisms, and the detritus of decaying cadavers (113–14). Far from the Romantic's empty abyss, the liquid ocean is therefore infused with life:

> Far as the eye can reach, you may see the water whitened with the marvellous abundance of the thick, fat, viscuous billows in which life is fermenting into new life. Over hundreds of square leagues it seems as though a volcano of teeming and fecund milk had burst forth and overwhelmed the sea.
>
> (107)

Constantly employing gendered metaphors that depict the ocean as a giant womb, Michelet defines sea life as enduring a permanent state of gestation:

> Their tenants seem, for the most part, fœtuses in the gelatinous stage, which absorb and produce the mucous substance, permeate and saturate with it all the waters, and give to them the fecund and nourishing powers of a vast womb, in whose depths an infinite succession of generations, perpetually float, as in warm milk.
>
> (118)

These metaphors depict the ocean as a global living organism that is in turn made up of other organisms, ranging from the microscopic polyp to the colossal monsters of the deep. The planet's oceans therefore become a giant organism constantly consuming, producing, mating with, birthing, and killing itself. This chaotic imagination of the evolutionary ocean's metamorphic, biological sublime offers perhaps the most abject obstacle to human identification with the sea.

Yet Michelet's application of the organism, rather than the human, metaphor to the sea offers a significantly new way of understanding the ocean. Although Michelet's use of the organism metaphor follows the discursive shift Foucault identifies in *The Order of Things*, which re-conceived nature as systems and phenomena produced by history and structured by internal laws, in Michelet's case the organism metaphor becomes a much more complicated reference.[12] Paule Petitier's Foucauldian study of Michelet in *The Geography of Michelet* (*La Géographie de Michelet*) suggests that the organism metaphor allowed Michelet to understand nature better by superimposing onto it the structures and timescales of the gestating body. Petitier argues that the organism metaphor finally rendered nature legible to Michelet:

> If we consider the development of an organism, each moment corresponds to a different form, a new inscription in space, and vice versa: each form that a being adopts over the course of its development can be precisely dated. This establishes an account of the relationship between time and space in the logic of the organism. . . . To prove the equivalence between the French territory and an organism is thus to assure the possibility to transfer time and space, history and geography.[13]

Here Petitier specifically refers to embryological theory, which helps to specify Michelet's understanding of the organism. However neither Petitier nor Foucault acknowledge the way that marine biology also destabilized the concept of the organism. Creatures such as the coral polyp, which naturalists struggled to classify as animal or plant; the hermaphroditic barnacles that challenged Darwin's understanding of reproduction; or the shape-shifting octopus that haunted popular literature all challenged nineteenth-century science's understanding of life. Whereas the Romantics' comparisons of the ocean to the human organism certainly made the sea more legible, Michelet's fantastic descriptions of marine life suggest that the organism metaphor may not be as clearly didactic as Petitier suggests. When Michelet applies the organism metaphor to the sea he does so to evoke its deep evolutionary history, its mysterious metamorphic forms, enigmatic biological cycles, and the other unknowns that marine biology had only just begun to explore.

As much as Michelet waxes fantastic about the strange biology of marine life, he also cultivates a discourse of the evolutionary ocean that insists on its melancholic nature. This pre-Freudian discourse of the melancholy depends on a progressive evolutionary logic in which the ocean and its primitive organisms long for the complexity of terrestrial life that they will never achieve:

> Great mother that hath commenced life, thou canst not perfect it; allow thy daughter, the Earth, to continue the work. You see it, even in your bosom; your children think of Earth and long for its fixity; they approach her, offer her their homage. It is for thee still to commence the series of new beings, by an unexpected prodigy, a grandiose rough draft of the warm amorous life, of blood, of milk, of tenderness which will have its development in the terrestrial races.
>
> (224)

Michelet's panache for exaggerated rhetoric and spectacular biology suggests that nature does not inherently follow the orderly evolutionary chain that his natural history presents. Indeed, *The Genesis of the Sea* catalogues the monstrous results, horrible hybrids, and evolutionary failures of the ocean's attempts to produce the human. Michelet blames these failures specifically on the marine environment, where the contaminating presence of its own sunken history prevents the ocean from ever attaining a more evolved state. Marine history is never buried, Michelet explains, but rather melancholically recycles in the deep:

> We know in reality no more about the composition of water than we know about that of blood. What we best know and can most safely affirm about the *mucus* of sea water, is that it is at once an Alpha and an Omega, a beginning and an end. Is it the result of the numberless deaths which furnish forth materials for new lives? No doubt, that is the general law; but in the case of the sea, that world of rapid absorption, the majority of the

> creatures there are absorbed while in full life; they do not slowly linger on towards death, as we on land do.
>
> (117)

This living liquid composed of sexual secretions and decaying matter blurs the distinctions between living and dead, present and past. Organic matter cycles through the waves and the organisms that compose the submarine environment. Rather than offering the Romantic figures of the ruined shipwreck or the cyclical shoreline as a metaphor for the ocean's history, Michelet's evolutionary ocean offers the image of a melancholy marine sink in which history is dissolved into the living water itself. This melancholy history troubles the otherwise seemingly infinite capacity for marine metamorphosis and biodiversity. Although Michelet aestheticizes the dazzling diversity and plasticity of marine life, through his evocation of the melancholy marine, this metamorphosis becomes associated with a deep history of primitive, abject, and unclassifiable forms.

The melancholy of Michelet's ocean is also heavily influenced by his gendered rhetoric, which portrays the ocean as the mother of all life, whose greatest weakness is her sublime fertility: "Such is the Sea, such the great *Female of the Globe*, whose ceaseless yearning, whose permanent conception, whose production and reproduction, never end" (113). Michelet's ocean therefore absorbs all the nineteenth-century's misogynistic associations of the hysteric, decadent, and the primitive with the female and the natural world.[14] Even though the ocean's fertility founded life on earth, Michelet describes it as a handicap, proposing that if she could only channel her endless sexual drive into something more organized than constant, chaotic rutting, the ocean might be able to evolve less monstrous life forms. The sea's melancholy is therefore partly caused by its inability to overcome the primitive instinct to reproduce:

> What is melancholy in the sea is not her carelessness to multiply death, but her impotence to reconcile progress with the excess of movement. She is a hundred times, a thousand times richer, and more rapidly fecund than the earth. . . .The sea is no other than the parturient and laboring womb of the globe. Her sole obstacle is in the rapidity of her births; her inferiority appears in the difficulty, which, so rich in generation, she finds in organizing Love.
>
> (221)

Whereas the Romantics saw their emotional turmoil and poetic grandeur reflected in the marine depths, Michelet sees hysterical biology, destructive sexual drives, melancholic longings – all of which prevent the ocean from attaining humanity. The mystery of how mankind ever emerged from his melancholy marine mother haunts *The Sea* from start to finish.

Within Michelet's melancholy history, one class of animals stands out: the marine mammal, which serves as the pivotal transition from marine to terrestrial life. The whale's mammalian physiology, complex family dynamics, and long historical relationships with maritime communities all demonstrate anthropomorphic qualities that allow Michelet to depict the whale as a sentimental and even pathetic figure. Much like the human, Michelet defines the whale as an exception to marine evolutionary history: "Not the slightest connection between this gentle race of mammiferæ, which, like our own, have milk and red blood, and the monsters of an earlier age, – horrible abortions of the primitive mud!" (228). However whales' immense size and weight condemn them to live within the violent seas that suit neither their respiratory systems nor their gentle character. As previously suggested, the whale arrives as an evolutionary aberration, ill-fit to its environment, and therefore quickly engenders the more moderate seal. Longing for the stability, society, and peace of earth, Michelet's maternal whale aptly becomes a metaphor for the sea itself, embodying the melancholic relationships to evolution, environment, and history that characterize Michelet's evolutionary ocean.

Michelet's maternal whale offers a striking contrast to Melville's monstrous sperm whale in *Moby Dick*, published only ten years before, and ignites a significant turning point for Michelet's sympathetic identification with sea life. Indeed, the third book of *The Sea, The Conquest of the Sea* (*La Conquète de la mer*) goes so far as to call for an international moratorium on whaling in order for global populations to stabilize. Michelet's sentimental depiction of the whale demonstrates how evolutionary theory allowed nineteenth-century readers to anthropomorphize marine biology in a new way, giving rise to an early conservationism. Moreover, Michelet remains at least partially interested in preserving whales because destroying them in turn destroys our chance to understand our own evolutionary history. Humanity's wanton destruction of the most highly-evolved marine organisms threatens his project of natural history and the progress of modern science.

The whale's lesson of conservation is further underscored by the mythical organism that appears as the last link in the chain of being between marine and man, a descendant of the whale that Michelet claims has already gone extinct. After describing evolutionary history from the first slimes to gentle giants, Michelet closes his narrative with the ocean's crowning anthropomorphic achievement: the siren, half-fish, half-man. "If we might give credence to certain traditions, progress did marvellously continue. The developed amphibious creatures, according to those traditions, approached nearer and nearer to the human form and became Tritons and Syrens, men and women of the Sea" (243). After such an exhaustive account of modern marine biology, Michelet's sudden leap to the siren would startle even the naivest reader. Yet in order to complete the chain of being Michelet needs some final link between the marine and man. Perhaps conveniently, he claims all trace of the siren has been lost and blames man, rather than the violent sea, for the siren's disappearance. Repulsed by their humanoid appearance, he explains, humans drove the hybrids to extinction:

> Those Syrens, too analogous to humanity, were all the more taken and detested for diabolic mockery. In such horror and hate were they held in

> the eyes of the middle ages that their appearance was considered a prodigy, an omen that God permitted to terrify sinners.
>
> (245)

Such grotesque imitation of man could not be tolerated: mankind therefore destroyed the species in order to maintain a definitive separation between the terrestrial and the marine, the human and the non-human. For Michelet this violent relationship with the ocean and its anthropomorphic evolutionary drive becomes definitive for mankind, especially in the modern industrial era, and chillingly predicts the Anthropocene.

Although Michelet is coy about the transgressive couplings and evolutionary aberrations that may have led from the siren to the human, he is certain of mankind's hostile reaction to the anthropomorphic marine. The violent arrival of humanity announces a new realm of nature, one in which mankind threatens the natural world into which he is born: "This creature, this tyrannical sovereign, can create a second nature within Nature. But what has he done with the first, with his mother, and his nurse? With the very teeth that she has given him, he has cruelly gnawed her bosom!" (247). To this extent, Michelet's account of the siren suggests that mankind is just another violent monster produced by the sea. Ending the evolutionary history of marine life with the first case of human-caused extinction also reframes Michelet's project of natural history as a melancholy modern project. Without the siren, humanity's closest evolutionary link to the marine, mankind will never fully know his evolutionary history. Although ending *The Genesis of the Sea* with the siren fulfills Michelet's project of connecting man to his marine lineage, it also ultimately condemns this connection as lost in history, and seals Michelet's natural history of the ocean as a modern melancholic myth.

Comparing Michelet's account of the siren to earlier Romantic reflections on the sea shows that to apply the anthropomorphic perspective to the ocean does not always result in a more legible marine. One might read Michelet's resort to the siren as the type of anthropomorphic rationalization that Adorno and Horkheimer claim to be the principle of the Enlightenment.[15] One might also see it as a desperate Romantic attempt to enforce universal natural law onto an otherwise illegible natural world. Although the appearance of the siren in *The Sea* renders Michelet's evolutionary history of the ocean arguably rational by completing the chain of being, it also breaks the text's scientific logic to embrace an utterly fantastic marine. The siren therefore models the way that Michelet uses marine biology to complicate imaginatively and anxiously his understanding of evolution throughout *The Sea*. Rather than rationalizing the pre-history of humanity through simple anthropomorphism, Michelet actively attempts to understand the non-human in human pre-history, to complicate his understanding of humanity, and to imagine the marine biology that produced us rather than praise the modern technological and scientific advances that promised to produce a coherent understanding of the marine. A far cry from the Romantic's easy identification with the empty abyss, *The Sea* shows how evolutionary marine biology challenged scientists, natural historians, and popular audiences to contemplate their evolutionary relationship to the sea.

Michelet's work may not resemble the scientific texts most remembered from the nineteenth century, yet it uniquely illustrates the discursive shift between the Romantic ocean and the evolutionary ocean as well as the provocative imagination and metamorphic submarine aesthetics produced by the co-evolving discourses of evolution and marine biology. Before these mid-century developments, a Romantic might look at the waves, admire the sublimity of its ancient scale, and compare it to the soaring ideals and tortured depths of the human soul. In the age of Darwin, the evolutionist looked at the waves and saw them crawling with life. This evolutionary ocean caused man not only to wonder at the sublimity of the sea's biodiversity but also to shudder at how his own evolutionary history might relate to the living waves. Throughout *The Sea* Michelet struggles with the grotesque, fantastic, and violent human portrait he discovers amongst the weird physiologies, waves of fecundity, and sublime violence of the sea. Nevertheless, he insists on the critical role of the marine environment in understanding man's history and future place in nature, and therein ushers forward a new era of marine exploration.

Notes

1 Natascha Adamowsky, *The Mysterious Science of the Sea, 1775–1943* (London: Routledge, 2016). Alain Corbin, *The Lure of the Sea: The Discovery of the Seaside in the Western World 1750–1840*, trans. Jocelyn Phelps (Cambridge: Polity Press, 1994). Judith Hamera, *Parlor Ponds: The Cultural Work of the American Home Aquarium, 1850–1970* (Ann Arbor: University of Michigan Press, 2011).
2 Corbin, *The Lure of the Sea*, 163–4.
3 Victor Hugo, *Oeuvres Complètes de Victor Hugo* (Paris: Eugene Renduel, 1835), 234–5. Translation my own.
4 Charles Baudelaire, *Les Fleurs du mal* (Paris: Poulet-Malassis et De Broise, 1857), 40. Translation my own.
5 Corbin, *The Lure of the Sea*, 127.
6 Jules Michelet, *The Sea* (New York: Rudd and Carleton, 1861), 141. All subsequent references to *The Sea* will be cited using page numbers within the text.
7 Lionel Gossman, "Michelet and Natural History: The Alibi of Nature," *Proceedings of the American Philosophical Society* 145, no. 3 (2001): 283–333.
8 Gossman, "Michelet and Natural History," 309.
9 Edward Kaplan, "Michelet évolutionniste," *Romantisme* 10 (1975): 111–28, Persée, doi:10.3406/roman.1975.5003, 119. Translation my own.
10 Jules Michelet, *La Mer*, 2nd ed. (Paris: Hachette, 1861), 251–4. Translation my own.
11 Adamowsky, *The Mysterious Science of the Sea*, 45.
12 Michel Foucault, *The Order of Things* (New York: Routledge, 1970), 139–40.
13 Paule Petitier, *La Géographie de Michelet* (Paris: Editions L'Harmattan, 1997), 153. Translation my own.
14 See Linda Orr's *Jules Michelet: Nature, History, and Language* (Ithaca, NY: Cornell University Press, 1976) for an extensive psychoanalytic reading of Michelet's maternal metaphors of the ocean.

15 Theodor Adorno, Max Horkheimer, and Robert Hullot-Kentor, "Odysseus or Myth and Enlightenment," *New German Critique*, no. 56 (1992): 109–41. doi:10.2307/488331. See Akita Lippitt's discussion of Horkheimer and Adorno in *Electric Animal: Toward a Rhetoric of Wildlife* (Minneapolis: University of Minnesota Press, 2000): "Horkheimer and Adorno argue that according to processes of Enlightenment understanding, the elements of life that evade comprehension or rational analysis are transformed into mythical figures that are then appropriated into thought through anthropomorphic transfiguration. In this way, reason manages to recuperate, or 'disenchant,' even irrational and supernatural phenomena. . . . For Horkheimer and Adorno, the dialectical process of reintroducing the monstrous into the folds of humanity through anthropomorphosis defines the principle of humanism and the Enlightenment" (79).

Plate 1: Figure 1.1 Jacques de Gheyn II, *Design for a Garden Grotto*, c. 1620-5, paper, 22.8 x 74.2 cm. Courtesy of the British Museum.

Plate 2: Figure 2.3 Jean-Baptiste Janelle (de Jonge), *Medaillonportret van een vrouw*, 1741, terracotta, 31.3 x 17 x 14.2 cm. Courtesy of the Rijksmuseum.

Plate 3: Figure 2.4 Capodimonte Porcelain Factory after Giuseppe Gricci, basin and ewer, 1745-50, soft-paste porcelain soft-paste, gold, 34 x 11.1 cm & 27.3 x 15.7 cm. Courtesy of the British Museum.

Plate 4: Figure 4.1 William Blake, *Newton*, 1795-c. 1805, colour print, ink and watercolour on paper, 46 x 60 cm. Courtesy of Tate Images.

Plate 5: Figure 4.3 William Saville-Kent, "Great Barrier Reef Corals," in *The Great Barrier Reef of Australia* (London: W.H. Allen, 1893), chromolithograph VII.

Plate 6: Figure 4.5 Jonathan Lamb, photograph of sunset over One Tree Island, 2016.

Plate 7: Figure 4.6 Jonathan Lamb, watercolour of rain over One Tree Island, 2016.

Plate 8: Figure 5.2 Thomas Rowlandson, *Venus's Bathing, a Woman Swimming in the Sea at Margate*, c. 1800, coloured etching. Wellcome Collection.

Plate 9: Figure 5.3 Benjamin West, *The Bathing Place at Ramsgate*, c. 1788, oil on canvas, 35.6 x 44.5 cm. Yale Center for British Art, New Haven, Connecticut.

Plate 10: Figure 10.4 Georges Malkine, *Sirènes*, 1926, oil on canvas. © ADAGP, Paris, 2018.

Plate 11: Figure 10.8 Screen grab from Damien Hirst, *Treasures from the Wreck of the Unbelievable* (2017) depicting the "salvage" of a Grecian nude.

Plate 12: Figure 11.1 William Lionel Wyllie, *Davy Jones's Locker*, 1890, oil on canvas, 1025 mm x 1360 mm. Courtesy of the National Maritime Museum Greenwich.

Plate 13: Figure 11.2 Joseph Mallord William Turner, *Tintern Abbey: The Crossing and Chancel, Looking towards the East Window*, 1794, graphite and watercolour on paper, 359 x 250 mm.

Plate 14: Figure 11.3 Jacques-Yves Cousteau and Louis Malle, *The Silent World* (*Le Monde du silence*) (1956). Screenshot of canted shot of the Thistlegorm (c. 40:20).

Plate 15: Figure 11.4 Jacques-Yves Cousteau and Louis Malle, *The Silent World* (*Le Monde du silence*) (1956). Screenshot of fish looking out of a window (c. 43:21).

Plate 16: Figure 11.5 Jacques-Yves Cousteau and Louis Malle, *The Silent World* (*Le Monde du silence*) (1956). Screenshot of the bubbles from the breathing wreck (c. 43:37).

Plate 17: Figure 12.1 Craig Mayhew and Robert Simmon, NASA GSFC, *The Earth's City Lights*, 1995. NASA.

Plate 18: Figure 12.2 Blue whale skeleton "Hope," 2017. Courtesy of the Natural History Museum, London.

Plate 19: Figure 12.3 Sir Joseph Boehm, *Charles Darwin*, 1882. Courtesy of the Natural History Museum, London.

Plate 20: Figure 12.4 NASA IMAGE satellite, *Aurora Australis*, 2005. The Earth Observatory, NASA.

Plates 21–32: Figures 14.1–12 Janet Laurence, *Deep Breathing: Resuscitation for the Reef*, 2016, mixed media: scientific glass, cast resin, silicon tubing, silk thread, pigments, fluids, marine specimens. Courtesy of the Australian Museum, Sydney, and the Muséum national d'Histoire naturelle, Paris. Photographs by Benjamin Huie and Lena Galangau-Querat.

Plate 21: Figure 14.1 The obscenity of the coral collapse

Plate 22: Figure 14.2 The warp of water, the geology below

Plate 23: Figure 14.3 The bloody thread

Plate 24: Figure 14.4 Swelling lament

Plate 25: Figure 14.5 Conduits of colour

Plate 26: Figure 14.6 Blood/chlorophyll

Plate 27: Figure 14.7 The paediatric ward

Plate 28: Figure 14.8 Lost habitats

Plate 29: Figure 14.9 Spheres: the isolation ward

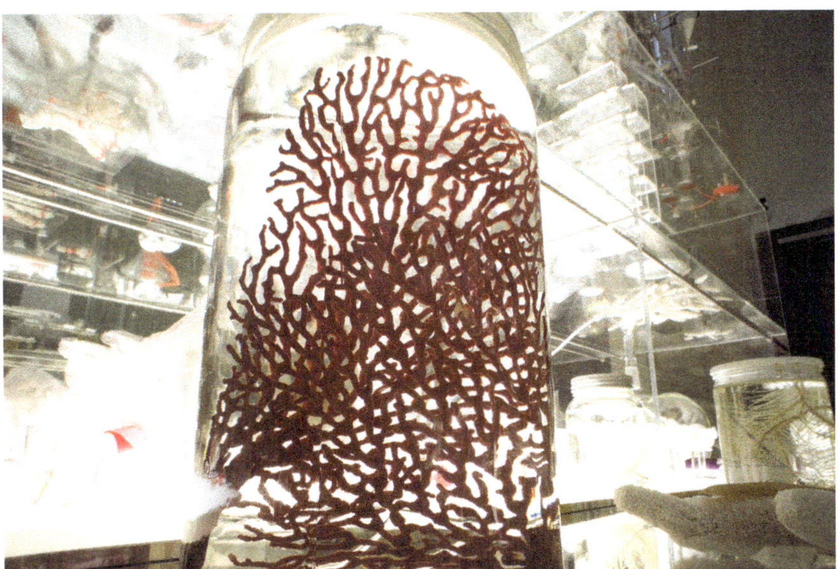

Plate 30: Figure 14.10 Geometry and geology

Plate 31: Figure 14.11 Intensive care

Plate 32: Figure 14.12 A turtle's view: emergency

7 "The Forsaken Merman," "The Little Mermaid," and early modernism

Undersea imagery for the dissociation and dissolution of culture

Samuel Baker

In an early poem, "The Forsaken Merman," Matthew Arnold develops a submarine fantasia that, in itself and in its relation to events on land, dramatizes a set of worries about culture: worries about its conditions of possibility, and about what can feel like its ultimate impossibility. Such anxieties about culture, I propose, animate much sea imagery from Romanticism onward, largely because it is in the wet picture language of a watery phenomenology that we find written many of the concepts of society, nature, psychology, and religion that matter most for the modern era. I won't try here to substantiate these claims with a full survey of such aquatic imagery and concepts, as they evolve from before Samuel Taylor Coleridge's "Rime of the Ancient Mariner" and J. G. Herder's "ocean of human fancies and dreams" through Virginia Woolf's *To the Lighthouse* and Sigmund Freud's "oceanic feeling" and thence up to the present day.[1] Instead, I retrace a few imagined submarine visions that capture, in dialectical images, structuring contradictions of modern, and modernist, culture. I sketch such images as I find them in Arnold's poem, and across an intertextual network that reaches back to Hans Christian Andersen's fairy tale "The Little Mermaid"; crosses over to William James's writings; and includes as other main nodes an essay and a poem by Arnold's successor as a water-obsessed, culture commanding poet-critic, T. S. Eliot. In this series of sketches, I especially seek to bring out the figures of dissociation and dissolution that surface in these authors' virtualized marine environments: main figures, I suggest, in the repertoire of tropes through which culture has been thought, felt, imagined, and otherwise experienced.

More basically, this essay asks, What makes us dream of mermaids? What is it about the aesthetics of the undersea that entices poets and philosophers to take imaginative recourse to their purported realm? Historical explanations for why oceanic experience became so imaginatively prominent across the early modern period can be found in other contributions to this volume and in various studies constitutive of the emergent fields of environmental humanities and, more particularly, maritime humanities.[2] This essay proposes a different sort of explanation for the salience of undersea aesthetics: it highlights the functionality of marine imagery for mediating thought about culture, thought about the nature of human society. Nothing on earth is more inhuman than the depths of the sea, given that no human there long survives; yet while this fact can make imagining the undersea seem an exercise in observing, and thence representing, pure nature, it can also make the undersea a convenient domain for imaginative social speculation. Remote from human experience, the undersea world serves well as a blank screen on which to project fantasias that reorganize and process the everyday materials of terrestrial life. The amphibious life of mermaids and mermen allegorizes the movement of thought across boundaries between the human and the inhuman, the social and the natural. The quasi-human embodiment of these creatures, meanwhile, is suggestive of the uncanniness attendant on the European experience of radical cultural difference from the peoples whom they encountered when they traversed the oceans.

Because the undersea world, far below the overseas territories of colonial expansion, gives modern explorers still further frontiers where their full panoply of instruments and institutions need to be mustered, what little sensory data we can derive from that realm garners a powerful aura. When writers observe the characteristic formal attributes of submarine aesthetics – phenomena like undulation, distortion, and slow motion; mutability, disjunction, and shifting perception; and those most attended to here, dissociation and dissolution – their observations register, exactly in the differences they chart from everyday terrestrial experience, the authenticity of the care with which the senses focused on the undersea, or the minds imagining it, attend to and communicate a natural environment. Art that registers such mindful care for nature accrues mediatory social power as culture. More broadly, across the range of human experience, the sensory immersion offered by undersea experience is special enough that we call its very form, exactly, "immersion." We name the wider category of experience in which we classify undersea experience with an aquatic metaphor. Like William James, whose *The Varieties of Religious Experience* – the subtitle of which, it bears emphasizing, is *A Study in Human Nature* – investigates "deeper plunges into mystical consciousness," we assimilate such total absorption to being, or to imagining being, immersed in water, often through ritual.[3] Undersea aesthetics does more than just provide a repertoire of metaphors for describing the zone where religion, culture, and social science can remake us whole again: it lends elemental form to some of the main ideas that modern culture develops of how bodies, souls, and selves are unmade, made, and remade.

At first glance, the undersea world of Matthew Arnold's poem "The Forsaken Merman" appears as a matter not of culture but of nature – albeit strange, fairy nature. Like the sea creatures Arnold depicts, readers of such verses as these can "bask in the brine,"[4] registering all the ways immersion in salt water affects the sensory environment:

> Sand-strewn caverns, cool and deep,
> Where the winds are all asleep;
> Where the spent lights quiver and gleam;
> Where the salt weed sways in the stream;

Where the sea-beasts, ranged all round,
Feed in the ooze of their pasture-ground;
Where the sea-snakes coil and twine,
Dry their mail and bask in the brine;
Where great whales come sailing by,
Sail and sail, with unshut eye,
Round the world for ever and aye.
(35–45)

These lines attend to minor oscillations, calibrating the minute differences that define the uncanniness of the submarine sensorium. How the sea bottom sand is strewn around; how lights quiver, and weeds sway, when the salt ocean, rather than the air above, is their medium: these fay stage effects are also natural facts the poem records. The poem likewise seems to register natural facts when it depicts the behavior – and in particular the undersea tempos – of sea-beasts, sea-snakes, and great whales, as each kind repeats its particular gestures, on its particular scale, without seeming to affect anything.

Yet while this tableau foregrounds the flickering, evanescent quality that inheres in perceptions of the undersea world at rest, it also conveys the potential the undersea world holds for constant work. It depicts an ordered, cultivated place. Its initial description, of the "winds . . . all asleep," is also its premise. Were this undersea world not asleep, but astir, were this realm disordered, its watery medium would become clouded. Clarification is itself a form of work, and the ordering hand of some master is implicated throughout the scene, preparing it on the plane of representation for its poetic apotheosis. Perhaps an Aeolus has lulled the winds; more surely, the "ranging" of the sea-beasts in "their pasture-ground" is the work of the Merman and his folk. The sea-snakes, as they "coil and twine" and "dry their mail," ornament the placid cavern with a controlled, threatening presence. It is as if they mirror, in the sea, the "mail clad-men" who when Branksome Hall is "idlesse all" nevertheless await, at the outset of Walter Scott's *The Lay of the Last Minstrel*, "the beck of the warders ten."[5] The whales, for their part, as they "sail" endlessly "round the world" "with unshut eye," shadow human merchant vessels, not to mention whaling vessels, that circulate untethered from natural cycles (44–5).

So the undersea household of the Merman – which is what is described here – is constituted by translated, but recognizable, human social practices, which is to say, it is a culture, in the sense of a cultivated zone, and moreover in the sense of a whole way of life, pastoral, georgic, epic, and even commercial. And just as it is the Merman himself who organizes this culture, it is also he who strives to adduce, to his household way of life, culture in the further sense of a medium of instruction and pleasure. The whole poem is presented as a monologue the Merman delivers to his children. (It is thus his own scrupulously clarified depiction of his undersea realm I have quoted.) Kinship, and the teaching of the young, is a further, perhaps a still deeper, layer of culture that the Merman is striving – and, as the poem opens, is failing – to organize in his realm. He fails, because he has been, and remains at the end of the poem, "Forsaken." He has been abandoned by Margaret, the human mother of his children, who now lack her care and instruction. Departing the undersea domestic sphere of the Merman's cave, Margaret has returned to land and ensconced herself in a church, whose bell rings out so loudly that its sound waves travel underwater and rouse the Merman to bring their children to pay her the fruitless visit that constitutes the poem's main action. Still, if in, and with, the poem the Merman himself fails with his project of acculturation and redemption, hope remains that the poem, its poet, and a faithful readership might succeed at redeeming the Merman, his mission, and perhaps even his wife for their own, literary culture.

What Arnold realizes, through the device of his Merman figure and the story of his alienation from his human wife, is the experience of traversing, and thence of capturing or failing to capture, extreme cultural difference. Arnold conveys the gain and the loss of transcultural experience with symbols that correlate with common bodily sensations of threshold-crossing mediation, especially sensations that characterize the human experience of the underwater environment. On the level of prosody, Arnold uses internal near-rhymes and subtly shifting and elongating diphthongs to convey the submarine resonance of the bell that reaches from the hill above to the sea cave below, and thence to us wherever we hear of it. "Down swung the sound of the far-off bell," he writes of the fateful peal that reaches Margaret and recalls her to her former life on land, to which she becomes again bound. When the bell thus takes on the distinctive timbre of submarine sound – a vital aspect of the poem's aesthetics of the undersea – and uses that auditory image to color terrestrial readerly experience, while opening spans of cultural immersion, it becomes an aural icon for the somatic reverie that gives rise to imagination. Yet the bell is also a call to prayer that as such threatens transcultural union. The swing of the bell as it brings sound, and, in the wake of sound, looks, and thence journeys, up and down, above and below water, commands for the poem the alternation of figure and ground, of culture and counterculture, that constitute its multistable image of literary culture as Christian or fairy, yet never both at once.

The bell in this way becomes a medium that symbolizes the experience of cultural mediation, of mediation both as it connects and as it holds apart. While it signals an opening, in so doing the bell also speaks of closure.[6] The medium of the book, by contrast, is a symbol for culture closed shut. After the bell has rung Margaret up to services and the Merman up in her wake, Arnold paints a scene that at first suggests sea and land might reconcile. Life above appears through a watery scrim when the Merman describes how he and the children rose from the sea

> To the little grey church on the windy hill.
> From the church came a murmur of folk at their prayers,
> But we stood without in the cold blowing airs.
> We climbed on the graves, on the stones worn with rains,
> And we gazed up the aisle through the small leaded panes.

She sate by the pillar; we saw her clear:
'Margaret, hist! come quick, we are here!
Dear heart,' I said, 'we are long alone;
The sea grows stormy, the little ones moan.'
But, ah, she gave me never a look,
For her eyes were sealed to the holy book!

(67–81)

Water may dissolve writing, while glass affords clear sight if not speech: but liquidity and its denizens, like fairy folk, reach the limit of their powers when they confront "the holy book." Ultimately, terrestrial humanity and Christianity repel Merman culture, except – and this exception seems the point of the exercise – insofar as the poem, with its lyric intensity, might lure young human listeners to stray from their books and perhaps even enlist as changelings in the Merman's brood, which must be a wholly imaginary affiliation, since actually to lower oneself to his court would be death. Arnold's poem perversely incites rebellion when it remains faithful to the Merman point of view, from which life on land appears as a counterculture of the "faithless." In his standard edition of the poem, Kenneth Allott suggests that Arnold's undersea setting is "an imitation of Byron's *Manfred* I i 76–87" (p. 99) and the Merman himself, damaged and plaintive, is suggestive of Arnold's penchant for Byronic renegades. Arnold's undersea realm is in an overdetermined way a man's world – quite unlike the undersea world of Hans Christian Andersen, with which it otherwise shares much.

"Cruel is she!" exclaims the Merman of Margaret at the end of the poem. "She left lonely for ever/The kings of the sea." The Merman styles himself and his band of brothers the benevolent patriarchs of a timeless province of kindness and imagination. When Arnold adopts their point of view for "The Forsaken Merman," in this as in a great deal else he reverses the polarities of Andersen's "The Little Mermaid," which had been published in English translation in 1846, just before Arnold wrote and published his poem.[7] Andersen's tale had followed a feminine (if not necessarily exactly female) heroine through a series of transformations that eventually entail her death. Like Arnold, Andersen narrates how his sea creature protagonist fails to secure a human love interest. In Andersen's well-known tale, a mermaid princess rescues a human prince from a shipwreck; falls in love with him; and trades her voice and tongue to a witch for human feet that can bear her, albeit with great pain, to join the prince on land (Figure 7.1).

The mermaid perishes of heartbreak after the prince passes her over for a human bride, finding however some measure of afterlife when "the daughters of the air" adopt her into their society of service. While on the basis of this plot, Andersen might well have titled his story "The Faithful Mermaid," his heroine cannot be said to be "forsaken." Andersen takes pains to specify the standard folktale cosmology in which humans have souls and mermaids do not, and the prince who passes over the devoted mermaid cannot be said to be cruel, or even faithless. The prince's choice simply represents the way of the human world. He is all too

Figure 7.1 John Leech, "The little Mermaid visits the Old Witch of the Sea," in *Bentley's Miscellany* Vol. XIX (1846) (English language debut of Hans Christian Andersen's "The Little Mermaid"), p. 376.

human, whereas Margaret's piety seems a species of inhumanity, if anything ironically confirmed by the glimpse Arnold gives of her dissociated longing for her "little Mermaiden" and her "golden hair" – for her undersea family's most fay feature. Meanwhile, unlike Arnold's Merman, whose extensive culture, perhaps above all his solicitude for his children, makes him virtually human, Andersen's mermaid is at least for the duration of his plot a merely natural creature, who does not possess, but only aspires to, the virtue of faith, and for that aspiration wins an extraordinary opportunity to labor for centuries to secure a soul, rather than simply dissolving, in spirit as well as in matter, into sea foam.

What has made Andersen's mermaid so memorable, by contrast with Arnold's now mostly forgotten Merman, is the palpable life afforded her by the Danish fabulist's inimitable fairy tale naturalism. Where Arnold's poem and its protagonists are bloodless and buttoned up, Andersen's tale is notoriously bloody and erotic. The tale degrades femininity, at times by seeming to exalt it, at other times by assimilating it to mere nature; but it also recognizes, if only through fantastic negations, the physical character of female experience. And Andersen resolutely embodies human experience more generally, not least when he gives undersea phenomenology a consequential force it lacks for Arnold. Whereas Arnold never explains how Margaret, the human love interest in his poem, could breathe under the sea, Andersen predicates his plot on the certainty that his human prince would drown were it not for the mermaid princess who saves him. As Andersen brings his various characters face to face with sex and death, he intensely associates the nature of the sea with the physical consequences of action. These associations converge in the tale's closing sequence, when the mermaid's sisters, shorn of their once-flowing hair, rise to present her with the weapon they've received in exchange for their locks: a dagger, with which the mermaid, at the last moment, refuses to spill the blood of the prince. This refusal precipitates the ultimate physical consequence for the mermaid: her dissolution, as she "once more fixed her dying eyes on him whom she loved better than her own soul, and plunged into the waves where her sweet body quickly melted away into foam." When Andersen affords his submarine, subhuman creature this death, the scene is at once more aquatic, more erotic, more vital, and more uncannily insubstantial than anything Arnold records in "The Forsaken Merman" – reminding us that for the Arnold of "The Buried Life," "the sea" is at bottom only where "life . . . goes" to be ultimately inaccessible.[8]

For Marina Warner, and for many of Andersen's critics, the "chilling message" of "The Little Mermaid" is that "cutting out your tongue is not enough." "To be saved," Warner writes, "more is required: self-obliteration, dissolution."[9] If Andersen's message is this nihilistic, it is also revolutionary, since when Anderson stages dissolution, he surfaces a recourse proper for the undersea subaltern whose inability to speak is intrinsic to her oppression. Because Andersen, as Jack Zipes astutely observes, understands the world in terms of an "essentialist ideology," his brutishly physical creatures require dissolution if they would assimilate in a more spiritual realm. Their only available rite of passage is their sublimation. Because Andersen's tale allegorizes failures of cultural translation that stem from fundamental inequalities of social situation, dissolution, as a figure for the realization of what is essential, paradoxically gives body to what Zipes calls Andersen's "discourse of the dominated," and does so strategically and, significantly, non-discursively.[10] Dissolution is the sea creature's avenue to the heavens above, especially as the practical relinquishment of the body corresponds to the self-abnegations of service and religion. As the daughters of the air tell the mermaid after her death: "thou, poor mermaiden, hast . . . loved and suffered, and therefore thou art raised into the world of airy spirits, where, in three hundred years, thou mayest win an undying soul." For Zipes, such hierarchies in Andersen mainly allegorize class ascent; no doubt they also involve colonial fantasies that encompass, and stem from, a range of social positions as they imaginatively revisit experiences of cross-cultural encounters in contact zones at the edges of European civilization. Andersen provides a quite conventional ethnographic description of the undersea kingdom of his mermaids at the outset of his grim tale. His moral about the stringent work requirements for acquiring a soul is as ethnocentric as it is evangelical; it makes the dissolution of the self a prime avenue, however paradoxically, for cultural advancement as well as religious salvation.

In "The Forsaken Merman," by contrast, Arnold writes ecumenically from the point of view of the Other; his poem, after all, is nothing more or less than his submariner's speech. Still, that speech's restriction to the Merman's voice reinscribes the grounds of its weakness, its inability to reconnect the dissociated worlds it traverses. As hard as the Merman strives to bind his children and Margaret back to himself by chains of poetic association, he is vanquished by the dissociation of his undersea kingdom from Margaret's abode on land, and by the mental state of dissociation from which Margaret now seems to suffer, having lost the ability to integrate her past experiences with her present. The very intensity with which the Merman struggles to reconstitute his family suggests how intimately such dissociation threatens them all. More broadly, Arnold suggests that both Christian religion and romantic imagination, having become separated, each individually lack the power to reintegrate: the poem demonstrates the limits of both discourses as culture and as faith. The poem can use the strangeness of its myth of undersea life to figure the gap between literary experience and religious experience, because doing so literalizes, and materializes, the elemental alterity of deep experience. The essential nature of life which for Andersen's tale achieves its apotheosis in dissolution, must for Arnold disintegrate through dissociation. As A. Dwight Culler observes, the dissociation Arnold tracks undersea is ambiguously social, cultural, or natural. Culler argues that Arnold's "poems which present an undersea world" depict love "thwarted by some social force": variously, "church, castle, and academic hall are responsible for the fragmentation of the modern world" he describes. Dissociation is the result of such thwarting of the social. Yet also, for Culler, "The Forsaken Merman" equivocates on whether dissociation is a social or a natural failure, when it suggests with its passively suffering children and its "imagery of waves, tides, and seasons" that "it could not have been otherwise."[11] And if Arnold equivocates here, his very equivocation, in the strength of its ambivalence, captures the dissociative energy that forcefully drives apart the fragments of modern experience.

Arnold's most celebrated marine figure for dissociative modern experience comes in another "Margaret" poem, of a kind: the later lyric "To Marguerite, Continued." This poem describes modern subjects "in the sea of life enisled/ With echoing straits between us thrown," doomed to "live *alone*" on their proper islands because "a God their severance ruled!/And bade betwixt their shores to be/The unplumbed, salt, estranging sea."[12] When he thus allegorizes the disassembly of the social, Arnold hearkens back to a locus classicus in Horace. In the third poem of his first set of Odes, that Roman poet expresses his appreciation for the daring exhibited by his contemporary Virgil by likening his fellow poet's audacity to a dangerous sea voyage. Horace then provides an initial, ancient instance of a modernism that registers the threat of dissociation with a maritime trope:

> Nequiquam deus abscidit
> prudens Oceano dissociabili
> Terras, si tamen impaie
> non tangenda rates transiliunt vada.

David Ferry renders these grammatically treacherous lines in an idiom compatible with a Christian cosmology; as a statement that "The purpose of the god who separated/One land from another land was thwarted/If impious men could nevertheless set out/To cross the waters forbidden to them to cross."[13] Ferry does so, one might suppose, as a modern translator of Horace, not just in terms of chronology but also in terms of aesthetic philosophy. He aligns divine purpose with literary achievement, albeit negatively, through a perfectly natural figure for dissociation as what good writing such as Virgil's avoids, through its cultural or religious, if not exactly its Christian, piety. Here, by contrast, is John Dryden's translation of these same lines from Horace:

> In vain did Natures wise command
> Divide the Waters from the Land,
> If daring Ships, and Men profane,
> Invade th'inviolable Main;
> Th' eternal Fences overleap,
> And pass at will the boundless deep.

The differences are subtle but significant. In Dryden's version, it is not God (Horace's "deus") but "Natures" (sic.) that "divide," and they do so "in vain," so long as men sail, which they will do.[14] While he envisages too-easy connection, not painful separation, Dryden resembles the despairing Arnold of "To Marguerite, Continued," in that he too offers no hope for a pious voyage of reassembly, for a quest sensible of the purpose of the sundering of the seas it navigates.

In verses like these, Dryden and Arnold instantiate, not just in style but also in substance, what T. S. Eliot would call "dissociation of sensibility." Nature, culture, society, and religion separate out in ways that may incite the rhetorical overleaping of boundaries, but that fail to realize the actual communion of different modes of experience, the transcendent experience to which poetry, Eliot thought, ought to testify. And if "The Forsaken Merman" likewise brings into conjunction sea and land, nature, culture, and society, and religion and romance, without however associating them properly, we can see on what grounds Eliot would dismiss that poem as one of what he calls Arnold's "charades."[15] For Eliot, drawing, I want to propose, on William James, such literature involves dissociation in the pathological sense the word had in contemporaneous psychology. According to Ian Hacking, "The word 'dissociation' entered English in 1890 thanks to William James," of whom, as is well known, Eliot made a close study. In November 1912, Eliot's latest biographer writes, the fledgling poet purchased his own copy of James's *The Varieties of Religious Experience.*[16] More than eighty years later, Eliot's various lectures on John Donne and his contemporaries would be collected under the title *The Varieties of Metaphysical Poetry*. When Eliot's posthumous editors echoed James's title, they signaled both the religious preoccupations of the main essay on metaphysical poetry Eliot published during his lifetime – the influential 1921 review in which he first adumbrates his notoriously vague concept of "dissociation of sensibility" – and the world-explaining ambition of that essay and of Eliot's subsequent elaborations of how such an event marked a watershed for the emergence of secular consciousness.[17]

For James, after Andersen and Arnold and before Eliot, dissociation is generally considered as a pathology, whereas dissolution, by contrast, is a dangerous but ultimately redemptive prospect, even though – especially because – it involves bodily experience beyond the limits of what conscious life can bear. Dissolution can even remedy dissociation. Along these lines, in the *Principles of Psychology*, the "hysteric" who suffers from a "'dissociated' and split off condition" can be "rescued" by "plunging her into the hypnotic trance."[18] Such plunging, elsewhere in James, is characteristic of a "dissolution" that is figured as marine experience and given an important role in the culture of mental life explored in *The Varieties of Religious Experience*. In that work, James explores how "the further limits of our being plunge ... into an altogether other dimension of existence from the sensible and merely 'understandable' world," a zone we might call "the mystical region, or the supernatural region."[19] This plunge would be fatal to our sensibility were it not for faith in a further truth: for James, here, both the scientific postulate and the religious faith that "God's existence is the guarantee of an ideal order." Granted this guarantee, in this zone "where God is," James writes,

> tragedy is only provisional and partial, and shipwreck and dissolution are not the absolutely final things. Only when this farther step of faith concerning God is taken, and remote objective consequences are predicted, does religion, as it seems to me, get wholly free from the first immediate subjective experience, and bring a real hypothesis into play. (361)

Dissolution, for James, brings us to the straits where science, or religion, or (traversing those zones of life) culture can undo dissociation by reassembling selves and society on grounds that are at once empirical and divine.

For Andersen, the dissolution of the little mermaid figures the achievement of such transcendence; for Eliot, such transcendence requires the unification of culture with science and religion that is the mission of the modern poet. In the essay on "The Metaphysical Poets" where he introduces his dire diagnosis of "dissociation of sensibility," Eliot prescribes as an antidote that "Poets in our civilization, as it exists at present, must be difficult," since "our civilization comprehends great variety and complexity," requiring the poet to "become more and more comprehensive, more allusive, more indirect, in order to force, to dislocate if necessary, language into his meaning."[20] Eliot's review essay about then-obscure seventeenth century poets can be a manifesto for modernism because the modern "method" it proposes turns out to be "curiously similar to that of the 'metaphysical poets'" (65), who "possessed a mechanism of sensibility which could devour any kind of experience." It was this "mechanism of sensibility," of the unification of thought and feeling, that to Eliot's mind was ruined when, "in the seventeenth century a dissociation of sensibility set in." Perhaps unsurprisingly, when Eliot, in his review essay, first quotes a metaphysical poem at length – citing a stanza from Donne's "A Valediction: of Weeping" – the stanza he quotes gives an allegory of fabrication that not only demonstrates, but also gives a marine figure for, the maintenance of the sensibility Eliot recommends. The quotation features elaborate figures, rapid association of thought, comprehensive reach – and worlds of salt water. The reach of this stanza is at once intimate and global, and its mode of association comes with its aggregation into all the waters of the world of the tears into which speaker and addressee alike dissolve:

> On a round ball
> A workeman that hath copies by, can lay
> An Europe, Afrique, and an Asia,
> And quickly make that, which was nothing, All,
> So doth each teare,
>
> Which thee doth weare,
> A globe, yea world by that impression grow,
> Till thy tears mixt with mine doe overflow
> This world, by waters sent from thee, my heaven dissolved so.
> (60)

For Donne, in these lines, and for Eliot, who quotes them so approvingly, as for James and Andersen – but not for Arnold, if his poem, as Eliot claims, only mimics the authentic process – dissolution is the figure for an extension of sensibility that is truly intersubjective because truly material (in this instance not just sentimental, but sensuous or even erotic). It is dissolution that assembles, at a most elemental level, the sundered continents (or islands, elsewhere in Donne) of experience whose unfortunate dissociation is what occasions the literary plunge through their watery medium.

Returning, in conclusion, to our jumping-off point for this tour of marine figures for cultural practice, Arnold's "The Forsaken Merman," we can note that, pace Eliot's dismissal of Arnold's craft, other critics have contended of the Victorian sage's work that "this poem is best seen in Modernist terms," since "paradox and ambiguity are the effects intended."[21] For Norman Friedman, "recreating a Danish legend for the purpose, Arnold has found his objective correlative," discovering "life under the sea is a very good symbol for the repressed life of the unconscious, with its barely suggested erotic overtones." Friedman goes on to quote the concluding lines of "The Love Song of J. Alfred Prufrock" for their resonance with Arnold's verses. Commentators on Eliot's early masterpiece, hunting sources for its concluding visions of "mermaids" and "chambers of the sea," invariably cite "The Forsaken Merman" as a forerunner.[22] Where is the dissociation, and the reparative dissolution, at work in "Prufrock"? Might that poem, which so famously ironizes the thoughts and feelings of its mediocre monologist, uniting them most at moments when the drama of faltering poetic voice becomes most overwhelming, provide a way to understand "The Forsaken Merman" as in like manner an achieved modern work? Or perhaps a contrast between these poems point up a failure in Arnold's early poetry: a flaw commensurate with the failure that Eliot would, decades after publishing "Prufrock," diagnose in Arnold's prose, calling out Arnold's "facile assumption of a relationship between Culture and religion" where "Culture is the ultimate value."[23] Is such a valuation of culture over religion incipient in Arnold's provision to his reader of the Merman's experience of a culture forsaken by the religious?

We've already seen how Arnold dissociates culture and religion in his poem, although we needn't follow Eliot in viewing such dissociation as a personal or political pathology, a vice, or a failure of artistic sensibility. If ultimately, for Eliot, dissociation is above all an occasion for redemption, one wonders if it was ever truly a crisis, more than just a drama. Perhaps when, with "Prufrock," Eliot outdoes Arnold, he does so by staging a pathetic charade of his own, redeemed only by the sensibility of its highly cultured protagonist's interior monologue. Concluding the poem, Eliot conjures marine, and ultimately submarine, scenes whose undersea aesthetics are proper, as we have been seeing, to extreme experience of the sort that can ratify such a sensibility's reach. These scenes overflow with the violent naturalism that was manifest in Andersen's mermaid tale, and that was latent, if powerfully so, in the lament of Arnold's merman. Eliot associates erotic fantasy with the most mundane of cognitions when he juxtaposes disparate imagery of combing hair and flowing waves; of liquid values of blood, "red and brown," and print, "black and white":

I have heard the mermaids singing, each to each.

I do not think that they will sing to me.

I have seen them riding seaward on the waves
Combing the white hair of the waves blown back
When the wind blows the water white and black.

We have lingered in the chambers of the sea
By sea-girls wreathed with seaweed red and brown
Till human voices wake us, and we drown.[24]

Above all, perhaps, Eliot develops a sense that immersion in material lyric makes a medium for the voice, poising such mediation as a possible answer dissolution might continue to make to dissociation. The mermaid will not sing to the speaker; when "human voices wake us," then "we drown." So, across this long early modernist moment of undersea aesthetics, the argument is made that it is only by drowning in an inhuman sea of faith – be its creed literary or Christian – that we can become our best selves, and think, feel, and comprehend the whole, terraqueous globe and its voices both above and below. Against this orthodoxy, a continued renegade life for Arnold's youthful fantasy "The Forsaken Merman" might indeed feel like a righteous rebellion, one no less in accord with whatever an aesthetics of the undersea might promise.

Notes

1 For a treatment of the first part of this evolution, see Samuel Baker, *Written on the Water: British Romanticism and the Maritime Empire of Culture* (Charlottesville: University. of Virginia Press, 2010).

2 *The Sea and Nineteenth-Century Anglophone Literary Culture*, eds. Steve Mentz and Martha Elena Rojas (New York: Routledge, 2017) offers a good overview of scholarship in these areas as it bears particularly on the literature covered in this essay. For a seminal treatment of the emergent importance of the marine environment for modern aesthetic consciousness, see Alain Corbin, *The Lure of the Sea: The Discovery of the Seaside in the Western World, 1750–1840*, trans. Jocelyn Phelps (Berkeley: University of California Press, 1994).

3 William James, *The Varieties of Religious Experience: A Study in Human Nature* (1902; New York: Routledge, 2008), 270.

4 Matthew Arnold, "The Forsaken Merman," in *Complete Poems*, ed. Kenneth Allott (New York: Longman, 1979), 96–100, l. 42. Henceforth cited parenthetically by line number except where a page number is indicated.

5 Walter Scott, "The Lay of the Last Minstrel," in *The Poetical Works of Sir Walter Scott*, ed. J. Logie Robertson (New York: Oxford University Press, 1957), 3.

6 Compare Andersen's tale "The Bell," in which two children, a prince and a pauper, wander through a forest in search of the source of a mysterious ringing, eventually meeting on a summit to enjoy both the sound of the bell and a view of the sea.

7 H. C. Andersen, "The Mermaid," trans. Lucie, Lady Duff Gordon, *Bentley's Miscellany* Vol. XIX (London: Richard Bentley, 1846), 377–90. Source hunters

seeking to lay bare the origins of Arnold's 1849 poem have traced them back to Danish folklore by way of relatively obscure passages in Andersen (see Allott's summary in his *Collected Poems* headnote, 96)—all the more reason to suppose that the more general atmosphere of undersea ennui Arnold conjures may owe a debt to Andersen's famous mermaid tale.

8 Arnold, *Complete Poems*, 271–6, ll. 97–8.

9 Marina Warner, *From the Beast to the Blonde: On Fairy Tales and their Tellers* (New York: Farrar, Straus and Giroux, 1994), 398.

10 See the chapter so named in Jack Zipes, *Hans Christian Andersen: The Misunderstood Storyteller* (New York: Routledge, 2005), 47–76; originally published in his *Fairy Tales and the Art of Subversion: The Classical Genre for Children and the Process of Civilization* (1983; New York: Taylor and Francis, 2006).

11 A. Dwight Culler, *Imaginative Reason: The Poetry of Matthew Arnold* (New Haven: Yale University Press, 1966), 21, 23.

12 Arnold, *Complete Poems*, 123–5; ll.1–4, 23–5.

13 Horace, *The Odes of Horace*, trans. David Ferry (New York: Farrar, Straus and Giroux, 1997), 10–11.

14 John Dryden, *The Works of John Dryden, Vol. 3: Poems 1685-1692*, eds. Earl Roy Miner and Vinton A. Dearing (Berkeley: University of California Press, 1969), 78, ll. 28-34.

15 T. S. Eliot, *The Use of Poetry and the Use of Criticism* (London: Faber and Faber, 1933), 105.

16 Robert Crawford, *Young Eliot* (New York: Farrar, Straus and Giroux, 2015), 178.

17 T. S. Eliot, *The Varieties of Metaphysical Poetry*, ed. R. S. Faber (London: Faber and Faber, 1993). For Eliot's elaborations of the "dissociation of sensibility" concept, see e.g. his essays on Milton, in *Selected Prose of T. S. Eliot*, ed. Frank Kermode (London: Faber and Faber, 1975), 258-74.

18 William James, *Principles of Psychology*, 2 vols., vol. 1 (New York: Henry Holt, 1890), 384–5. Elsewhere in this work, James describes a "law of dissociation by varying concomitants" (vol. I, p. 506, vol. II, p.345) in terms that relate the theory of dissociation he derives from Janet to associationist theories he cites from Hume and others.

19 James, *The Varieties of Religious Experience*, 360. Henceforth cited parenthetically by page number. See Joan Richardson, *A Natural History of Pragmatism* (New York: Cambridge University Press, 2007), for a highly relevant explanation of how James's work unfolds within a significantly aesthetic, if also scientific, understanding of humankind's natural environment.

20 T. S. Eliot, "The Metaphysical Poets," in *Selected Prose*, 59–67, 65. Henceforth cited parenthetically by page number.

21 Norman Friedman, "The Young Matthew Arnold 1847–1849: 'The Strayed Reveller' and 'The Forsaken Merman,'" *Victorian Poetry* 9, no. 4 (Winter, 1971), 405–28, 426.

22 See most recently *The Poems of T. S. Eliot: Volume I, Collected and Uncollected Poems*, eds. Christopher Ricks and Jim McCue. (Baltimore: Johns Hopkins University Press, 2015), 9, 398.

23 T. S. Eliot, "Notes Towards the Definition of Culture," in *Selected Prose of T. S. Eliot*, 292–305, 295. Capitalization of "Culture" follows the original.

24 T. S. Eliot, "The Love Song of J. Alfred Prufrock," in *Collected Poems 1909–1962* (New York: Harcourt, Brace & World, 1970), 7.

8 Encountering living corals

A nineteenth-century scientist and artist reveals the underwater realm

Iain McCalman

Dr John "Charlie" Veron, a young newly-minted PhD, was in 1972 appointed to James Cook University in Townsville, Queensland, to study Great Barrier Reef corals – the first full-time Australian research biologist to do so. That it should have taken Australia so long to explore the biology and ecology of the largest coral reef formation on the planet comes as a shock, particularly given that the Reef would within a decade be awarded World Heritage status as "the most impressive marine area in the world."

Part of the reason for this scientific tardiness was technological; it was only in the 1960s that open-circuit and rebreather scuba apparatus became cheap and reliable enough for students like Charlie to hire and use. It is a truism that without the capacity to breathe underwater for substantial periods, the ability of scientists to observe living corals and their habitats was extremely limited. In fact, Charlie, whose PhD qualifications were in etymology, was the sole applicant for the position because the advertisement had stipulated the need for scuba diving qualifications and experience. Luckily, he'd belonged to a student scuba club for several years and once undertaken a holiday dive with friends off South Solitary Island on the north coast of New South Wales. Here he had discovered and made a rough survey of a hitherto unknown embayment of tropical corals, the beginning of what was virtually a self-taught career in marine biology.

Employed by the university to map and describe the multiplicity of Great Barrier Reef corals, Charlie was sent first to visit the great natural history museums of London, Paris, Berlin and New York in order to examine the "type specimens" that underpinned existing coral taxonomies. To his chagrin, he found that a great number of these supposed "types" were not in fact distinct species. Desk-bound taxonomists who'd never seen living corals underwater had named as "new" species what were in fact local variations that were easily tracked underwater. He later estimated that five thousand different names had been given to what were actually only two thousand genuine species. The major offender was a legendary British pioneer of coral taxonomy, Henry Bernard. Charlie reflected wryly that if this great man had been able to spend even one day swimming underwater on the Great Barrier Reef, "his world would have been turned upside down."

Charlie Veron would over the next thirty years go on to become one of the greatest coral scientists in the world, spending more than 7000 hours underwater, observing and mapping Scleractinian corals both on the Great Barrier Reef and all over the globe. Having the ability to observe, study and photograph corals underwater enabled him to discover and describe some twenty percent of the world's reef-growing species and to win the Darwin Medal for his brilliantly original contributions to our understanding of coral evolution.[1]

In this chapter, however, I want to step backwards into the pre-scuba times of the mid-to-late-nineteenth century to explore how a marine zoologist, William Saville-Kent, pioneered the biological and ecological study of Great Barrier Reef corals without having the capacity to move, breathe, and photograph underwater. Indeed, like most of his scientific contemporaries, Saville-Kent could not swim at all, yet he was able to develop ingenious ways of offsetting this handicap so as to penetrate the underwater realm from dry land.

William Kent was born in Sidmouth, Devon, in 1845, the youngest child of a large family that would become still larger after his father Samuel's second marriage. William undertook his initial post-school education in geography and mathematics at King's College, London, courtesy of a death bequest from his artistic mother, who'd also inspired his lifelong love of painting. He seems to have gained his initial passion for natural history thanks in part to his slightly older sister Constance, who in 1859, at the age of thirteen, smuggled home from boarding school a copy of Charles Darwin's newly-published *On the Origin of Species*. Several years earlier, however, both William and Constance, aged 11 and 12 respectively, had attempted to run away to sea to escape a life of misery and humiliation inflicted on them by their hated stepmother, their former governess who had conducted a flagrant affair with their father while their ailing mother was still alive.

In 1860, Constance, assisted by William, had secretly murdered their three-year-old half-brother in an act of desperate vengeance against their stepmother. Lacking evidence, the police had been unable to charge anyone. Five years later Constance confessed to the crime in an act of adolescent martyrdom, taking sole responsibility for this sensational crime, at least in part to protect her beloved younger brother's scientific aspirations. The police remained convinced that she was protecting William, but again could marshal no proof. Though she was initially condemned to death, the sentence was later commuted to twenty years in prison. Throughout this sentence William ignored Constance's many pleas for him to visit or write, concentrating all his energies instead on building a scientific career. The buried trauma of this childhood crime also goes some way towards explaining his stiff and repressed adult personality.[2]

Despite having lived and played on the seaside during his boyhood, William Kent's introduction to the scientific study of corals required no saltwater experience. In 1868, at the age of 22, he was engaged to assist the Museum Conservator of the Royal College of Surgeons, Sir William Henry Flower, in arranging John Hunter's considerable invertebrate collection, which included extensive coral samples and fossils.[3] From there, John Gray, the Keeper of the British Museum's Zoology department, employed him in October 1868, "to name, arrange and make a manuscript catalogue" of their collection of stony corals known as *Madrepores*.

William later flattered his patron Flower that he'd first been "smitten" with corals at this time, but he seems in practice to have found the work dull and repetitive.[4] Most of it consisted of translating from French into English the descriptions of 900 coral species contained in *L'essai sur l'histoire naturelle des corallines*, a massive eighteenth century compendium of French and British corals. Though conscientious, Kent's jejune effort left his employers unimpressed. Even so, he was promoted in October 1869 to the lowly but permanent post of Junior Assistant in the British Museum's Geology department, where he was expected to continue working on the taxonomy of coral fossils.

At this time it was not uncommon for the scientific study of corals to fall under the province of geology. Though coral polyps themselves were recognised as animals, the massive underwater structures of limestone created by the skeletons of reef-growing corals attracted prime concern because of their threat to naval and merchant shipping. Charles Darwin's brilliant 1835–36 hypothesis on the origins of coral reefs through ocean floor subsidence had been made on the *HMS Beagle*'s naval survey, and had been achieved, as he later said, by using essentially deductive geological methods. He had studied Admiralty charts, examined coral fossils within the stratigraphy of elevated Argentine mountains, sketched bird's-eye views of South Seas reefs from the *Beagle*'s mastheads and spent a few weeks wading in the lagoon shallows and vaulting onto the fringing reefs of the Cocos-Keeling atolls in the Indian Ocean.[5]

Though employed as a museum geologist, William Kent was drawn increasingly during these years to the mode of marine zoology practised by Darwin's fiercest disciple, Thomas Henry Huxley, who held the Hunterian Chair at the Royal College of Surgeons from 1863 until 1869. Huxley had made his own early reputation studying the structure and behaviour of *Medusae* jellyfish and other pelagic creatures caught from the decks of the *HMS Rattlesnake* during the frigate's 1840s naval survey of Northern Australia. Attending Huxley's lectures and avidly reading his papers and books shifted William's focus away from making taxonomic descriptions of inert coral specimens and towards investigating the physiologies and behaviours of living marine creatures within their oceanic habitats.[6]

Under Huxley's influence, William Kent joined the Royal Microscopical Society in the late 1860s, where he was introduced to a new technology and set of methods that enabled him to explore the fascinating "new world" of marine underwater biology. Using microscopes William began to observe the forms and behaviours of *infusoria* – minute swimming marine organisms propelled by flagella. In 1869 he published his first scientific paper in the *Monthly Microscopy Journal* describing several new species of these unicellular animals, which he had discovered in seawater collected at the Victorian Docks in London. Soon he was elected a Fellow of the Society and would later establish a sufficient reputation to win three successive grants from The Royal Society to publish a massive three-volume *Manual of Infusoria* in 1880–2, dedicated to his patron, Thomas

Henry Huxley.[7] In it, thanks to the revelations of the microscope, he described hundreds of new species of infusoria, organised into genera and families.

In 1870, while still working doggedly on the inert coral specimens at the British Museum, William also experienced a transformative piece of luck when a wealthy Fellow of the Geological Society, Marshall Hall, invited him to join a research expedition to Portugal and Spain in his private yacht *Norna*. Given the rarity of such an opportunity, William was able to persuade the Museum to grant him leave to study protozoa and sponges off the coast of Portugal. Three weeks of collecting in Vigo Bay using dredges enthused him with the desire to work in the future as a zoologist, committed to collecting and studying the behaviour of marine creatures within their living habitats.

Inspired by this experience, he followed it in the same year by taking his annual holiday at Guernsey to explore with a group of zoological friends the normally underwater habitats of marine sponges and corals as these were exposed to view for a few hours during the extreme sea water retreats of low spring tides. Henceforth, observing, collecting and sketching at tidal pools during spring tides became a habitual part of his zoological repertoire, repeated in March the following year at St. Leonard-on-the-Sea, where his stint of collecting added new sponge species to the British Museum's collections.[8]

For a technologically gifted zoologist like William, the step from exploring tidal pools at the beachside to recreating them artificially within aquaria was not insuperable. Small household aquaria were already commonplace within middle class households after the 1850s. In 1872, the naturalist entrepreneur W. A. Lloyd, who had witnessed in Paris the commercial entertainment appeal of large-scale public aquaria using machine-driven water circulation systems, opened the first British equivalent at Crystal Palace in February 1871. From the outset, too, enterprising zoologists like Richard Owen and the ichthyologist Francis Buckland recognised the scientific potential of these facilities to study and replicate the lifecycles of depleting food fish stocks like salmon, oysters, herring and lobster.

Buckland, the Inspector of Salmon Fisheries from 1867, seems to have persuaded ambitious William Kent to abandon his ill-paid British Museum work in favor of taking up a succession of positions as an aquarium biologist, working first in Brighton, then at a variety of new aquaria in Manchester, Great Yarmouth and London. Here, during the 1870s and early 1880s, he combined his biological and engineering talents to become an expert in the aquaculture and acclimatisation of foreign species and also in improving water quality so as to keep marine animals alive through improved methods of water circulation and machine-driven aeration. Access to newly revealed dimensions of marine species' behaviour and habitat also enabled him to enhance his reputation by publishing short research papers in *Nature*, in which he reported pioneering observations on changes in the colour and behaviour of demersal fish and laid bare the complex sex life of the octopus. He also experimented successfully in new methods to overcome the nagging problem of preserving the colour and vitality of seaweeds by increasing their air exposure and intensified water circulation. Visitors could now watch entranced as live seaweeds danced underwater in alluring colours instead of lying like leached corpses on the strand. William Kent also cracked some of the longstanding mysteries of marine metamorphosis by hatching lobster eggs, monitoring the transformation of whitebait into herring, and working out the reproductive systems of cod, whiting and lobsters.[9]

In each new aquarium position, however, William Kent as biologist soon began to chafe against the way that the commercial managers gave priority to popular entertainment over the needs of scientific research. His prickly temper ensured, too, that these frictions usually culminated in public newspaper exchanges followed by William's outraged resignations. For several years in the early 1880s he pinned his hopes on obtaining government support for a private scheme to establish an aquarium and marine research centre on Jersey, and, when this failed to eventuate, on obtaining a position in a government-sponsored national research facility at Plymouth. It comes as no surprise, given his quarrelsome disposition, that he was passed over in favor of better-connected and probably more diplomatic marine biologists.

Just as his scientific career seemed stalled, William Kent's childhood past also threatened to catch up with him. As the time grew nearer for Constance to be released in 1885, at the end of her long imprisonment, William Kent had to face the possibility that his sister might seek revenge for his ingratitude towards her sacrifice. As someone whose waxwork figure featured in Madame Tussaud's as the most notorious murderess of the century, the publicity that would accompany her release also threatened his career. These potential crises were averted, however, when William, under the newly tweaked surname of Saville-Kent, accepted a position in 1886 as Inspector of Fisheries in Hobart, Tasmania. It is unclear, though, whether he did so in an effort to flee from Constance or as part of a plan to relocate her secretly along with the remainder of his siblings who also opted at this time to make new lives for themselves in Australia.[10]

In Tasmania it seemed for a time that William had at last achieved his dream of operating a marine research facility where he could experiment with new aquatic technologies to discover the life cycles of food fish and cultivate them artificially within simulated underwater habitats. Initially working in his own Hobart backyard, Saville-Kent built a small fish hatchery comprising eight wooden tanks with glass fronts, into which a hot air engine pumped oxygenated seawater from a central container.

This miniature facility became the pilot for a much larger hatchery and marine research station, which he then had erected at Battery Point on the edge of the Derwent River. Here he was able to use a 300-foot tidal river frontage to pump seawater directly into a hatchery reservoir. He also fenced off an acre of tidal flats, in which he dug a series of small, separated ponds, some of which carried experimental timber frames for oyster growth. Within this hybrid arrangement of aquarium and natural ocean pools he began cultivating both local and imported food fish species, as well as rock lobsters.[11]

His new Tasmanian paymasters, however, soon proved as hostile to marine research as the British commercial aquarium managers. An influential group of faux aristocrats on the Tasmanian Salmon Commission in Hobart who had lobbied to hire Saville-Kent wanted him to reciprocate by providing expert confirmation of their efforts to introduce the Royal British sporting fish, *Salmon salar*, into Tasmanian waters. This William refused to do. Instead, he pointed out bluntly in the local newspaper that the much-touted claims of local anglers to have caught royal salmon were actually nothing more than over-grown American trout. Introduced *Salmon salar*, he argued, had swum out to sea because they found the Tasmanian river waters too warm. The Salmon Commissioners responded by accusing him of being an impractical academic and then seized the opportunity to terminate his position at the end of 1887. During this difficult time, Constance, who – whether invited or otherwise – had followed her brothers and sisters to Australia under an assumed name in May 1886, seems also to have been living secretly with William and his wife in their Hobart house.[12]

In 1888, the year of Australia's centenary, William's luck suddenly changed. An unsolicited request from the Queensland government to report on the state's main oyster sites in Moreton Bay was followed by an invitation to accompany Captain HP Foley Vereker of the *HMS Myrmidon* on a survey voyage of the Cambridge Gulf on the north coast of Western Australia. "With alacrity," William grabbed at this chance to escape his troubles and explore a tropical region unknown to science.[13]

Even the transit voyage on the China Navigation Company's steamer, *Tsinan*, to join the Cambridge Gulf expedition in Darwin proved life-changing. He persuaded the ship's captain to call for a few hours at the Cairncross Islets on the Great Barrier Reef, around 70 kilometres north of modern Mackay. They arrived at dusk when the upper platform fringing reef was partially uncovered by the tide and William, who'd never seen a living reef, was stunned by the sight of corals "growing in their native seas and in their wonderful living tints."[14] Having only a few hours, he threw himself into a frenzy of collecting and drawing. He sketched black *beche de mer* (*holothuria*) pushing particles of sand and coral into their circular mouths; purple starfish thrusting their spinous arms "in every direction apparently seeking for food"; and semi-transparent pink *Synapta* floating ethereally in shallow tidal rock pools. "Unsolved mysteries" seemed to confront him wherever he looked: the unknown taxonomies and ecologies of the thousands of chambered spiral shells scattered on the reef; clusters of young living *Madrepore* stony corals floating from unknown locations on chunks of pumice; hordes of tiny oysters clinging to mangroves, and apparently new to science. Not surprisingly, the northern hemisphere zoologist fell instantly and permanently in love with this enticing glimpse of the underwater marine world of the Great Barrier Reef.[15]

Having to wait at the northern port of Darwin as the *HMS Myrmidon* underwent some preparatory repairs proved to be a second transformative experience. William passed the time by volunteering to help gather and catalogue a collection of local fish assembled by a Darwin policeman and amateur naturalist, Paul

Foelsche. Not only were these tropical species excitingly new and different, but Foelsche also turned out to be an enthusiastic amateur photographer who'd produced striking likenesses of the fish as they swam in shallow tidal pools.[16] Hearing William's later excited descriptions of Foelsche's photographic achievements, his wife Mary Ann gave him a "modest form of camera," which henceforth became an essential tool in his scientific armory.

News of the success of the Cambridge expedition impressed influential Queensland scientists, especially because William presented a subsequent paper discussing his discovery of several marine specimens new to science. These triumphs added weight to a proposal for a job as a fisheries inspector that William had dashed off to the Queensland Premier, Sir Thomas McIlwraith, just before embarking on the voyage. As it happened McIlwraith was concerned at the time about the grave depletion of the state's marine resources. In particular, the once lucrative pearl shell trade was bringing plunging returns because of overfishing and the incursions of Japanese luggers. New licensing regulations passed in 1881 had failed completely to arrest this decline. By March 1890, even the pearl fishermen of the Torres Straits were pressing for restrictions to be placed on the harvesting of immature pearl shell.[17]

In the same month the Queensland Premier offered William a three-year, full-time position as Commissioner of North Australian Fisheries on the substantial salary of £800 per year. He promptly traveled up to Brisbane by train, accompanied by his wife Mary Ann and another unnamed member of his family, who was almost certainly Constance. For the next twelve months or so, the threesome shared a house at Kangaroo Point overlooking the Brisbane River, likely offering the two siblings a chance to restore their broken trust. In the winter of 1890 Mary Ann and a female relative then travelled back down to Sydney by rail on an unspecified errand.[18]

Soon after this, Constance's movements can be tracked with some certainty. The year 1890 marked the beginning of her redemptive new career in Australia. Starting out as a volunteer in Melbourne's typhoid tents, "Emilie Kaye," as she now called herself, went on to enroll at the Alfred Hospital in Melbourne and from thence to embark on what would turn out to be a long and distinguished life as a nursing sister.

Fresh from the mire of politics in Tasmania, her brother was eager to show off his practical usefulness and economic value as a marine scientist and resource manager. A quick review of Queensland's fishing industries, plus conversations with McIlwraith, suggested an urgent need for what he called "a redemption" of both the edible oyster and pearl shell industries. He discovered that the Torres Strait pearling industry, which had been one of Queensland's leading revenue producers at around £70,000 a year, was now so exhausted that much of the harvested shell was too small even for button manufacturers to use.[19]

From the outset William determined to deploy his skills as an aquarist and expert on fish cultivation methods to create sustainable marine industries within the Great Barrier Reef. Nobody, however, had yet devised a way of cultivating pearl shell. Most of the region's shallow pearl-shell beds were exhausted; and the deep-sea beds were difficult to harvest and vulnerable to theft from foreign luggermen. Veteran Torres Strait pearlers were skeptical, however, of William's novel plans. They opposed his suggestion that immature pearl shell be transferred to manageable and protected shallow pools at Thursday Island on the grounds that this would entail severing the oysters' abysses (natural anchor cables), which would cause the young oysters to die or drift away on the currents. They even doubted whether the oysters would survive their transit voyages from the deep beds to new locations.

Between 1889 and 1891 William made three extended stays in the Torres Straits to tackle these problems and to prove his critics wrong. Enthusiastic support from two of the region's most influential Europeans made his task a good deal easier. Frank Jardine was a leading pearler-adventurer from Somerset near the Albany passage, and John Douglas was official Resident at Thursday Island in the Torres Straits. Both helped William to visit the deep water "Old Fields" near Badu Island, where, using suitable boats and skilled Islander free divers, he was able to collect abundant samples of immature deep-water shell.

Jardine also provided William with sites near his cliff-top house at Somerset on the tip of Cape York, where the biologist could undertake oyster transplantation experiments using improvised aquaria of giant clamshells filled with seawater. Having successfully transported young pearl oysters from the deep-water grounds of Badu by immersing them in these portable aquaria, William transplanted the immature shell in a series of shallow rock-pools near John Douglas's Thursday Island Residency that offered current-washed environments perfectly suited to oyster growth. To his delight, the young pearl oysters "adapted themselves with alacrity to the novel environment" by growing new abysses with which to anchor themselves until their shells were sufficiently heavy to resist the currents through their own weight. In one six-week period the oysters added an astonishing half an inch to their shells.[20]

Willliam also quickly discovered how little was known about the character and behaviour of the different coral types within the Great Barrier Reef. The celebrated expedition of 1872 of the British oceanic vessel, the *Challenger*, had collected sixty-one species of Australian reef corals, yet William, who made no claim to exhaustiveness, discovered over seventy new species of *Madrepores* alone.[21] By exploring in small boats, wading in reef shallows, peering into pools with a glass viewer, he was also able to map how varieties of different reef locations produced distinct combinations of coral types. *Madrepora* corals like the *Staghorn* flourished at some Port Denison reefs; luxuriant bush-like clumps of *Millepora* at some Palm Island sites; mushroom corals were common off Adolphus Island; and leathery bright green *Alcyonaria* corals dominated at most of Thursday Island's reefs. And on wave-crashed outer-barrier reefs like Warrior Reef he found no signs of the hummocky *Goiastrææ, Maendrinæ*, and *Astræceæ* corals that were present at most inshore and fringing reefs.[22]

Reef corals proved more protean than he'd expected. Colours often varied markedly amongst separate colony stocks of the same species and even within what were clearly different growth epochs of the same colony. This seemed to suggest that the corals were undergoing environmental adaptations in different depths and water conditions. Non-coral species, he noticed, also made a surprisingly important contribution to reef building. Seaweeds called *Nullipores* with lime-encrusted tissues, shells of minute *Protozoic* organisms called *Foraminifera*, and the skeletons of sea urchins, starfish and *trepang* all appeared to contribute to the lime cement and conglomerate that built the reefs. [23]

Knowing exactly how fast corals could grow also suddenly assumed a dramatic significance when, on 28 February 1890, the *HMS Quetta* – the pride of the British India and Australian Steam Navigation Company's fleet – hit a submerged coral pinnacle in a well-charted area between Albany and Mount Adolphus Islands. The huge steamship sank in three minutes, with the loss of around half of its 292 passengers and crew. Asked to investigate the site, William used Islander free divers to measure and map its underwater contours. He concluded that the coral pinnacle responsible had grown rapidly during the thirty years since the last survey to become an unmarked hazard; and he urged that accurate studies of coral growth rates be undertaken urgently.[24]

When it came to studying the prolific Reef order of *Actinarians* or anemones, William received unexpected help. Soon after arriving in the Straits he met Alfred Cort Haddon, a younger Cambridge scientist with a reputation for having studied British anemones. Initially, too, Haddon's mission seemed identical to his own:

> I propose to investigate the fauna, structure and mode of formation of the coral reefs in the Torres Straits ... to map the raised and submerged coral formations ... to investigate the fauna of the lagoons of the shore exposed at low tide and of the submarine slope ... to endeavor to determine the zones of different species of coral and of associated invertebrates, and also what conditions of light, temperature and currents are favourable or otherwise to the different species.[25]

As it turned out, however, the two men would not compete for long because Haddon developed what would become a life-long fascination with the anthropology of Torres Strait Islanders. In the meantime, the two Englishmen were happy to collaborate. Each named a new species of anemone after the other. William's discovery, a giant 24-inch anemone, *Discosoma Haddoni*, had the special attraction, he noted, of operating a "commensal" relationship with a small coloured clown fish and a pink-striped prawn. He speculated that the fishy visitors paid for their safe haven by serving as lures to attract other marine creatures to follow them into the anemone's mouth.[26]

William soon came to believe, too, that he'd acquired an advantage over all contemporary marine biologists by having taught himself to use photography as a tool for the scientific study of reefs and their inhabitants. In a pre-scuba world, living corals were rarely seen because of their underwater growth. As he wrote to his former patron, Sir William Flower, "such photos of the coral animals ... have not previously been taken and they will ... illustrate what valuable science can be rendered by the camera in illustration of the living aspects of these organisms."[27]

Photographing these clumps of living coral was possible only during their fleeting periods of exposure at low spring tides. Then, however, and provided one worked with great speed, they could, William boasted, "be reproduced with the fidelity that photography alone can compass," and that no pencil could equal.[28] *The Brisbane Courier* would later claim that he'd been able by this means to bring to light coral thickets that even local fishermen had never seen.[29] His photography also enabled him to record the geological structures of reefs, map the distribution and relationships of reef corals, and capture the exact likenesses of marine species while they were still brimming with life. Photographs could also serve as precise baselines for measuring changes in coral growth and distribution over time.

To achieve all this, a scientist-photographer needed great patience and physical stamina, as well as technological virtuosity. Because specialist equipment for scientific photography was non-existent, William had to devise his own square lens frame, as well as to build an extra supporting leg on his tripod in order to take corals and tiny crustaceans from a vertical position. He also experimented endlessly with using different lens types to find the most suitable focal distances for capturing the true size of his specimens. Wherever possible, he photographed marine creatures in his portable shell aquaria so as to replicate original environmental conditions and retain true natural appearances.[30]

Capturing physical exactitude was not his only concern. The beauty and romance of this "fairy land of fact" appealed to the artistic side of his personality. Lean, bearded and angular, wearing a solar hat and a trim fitting suit, William cut a romantic figure as he waded through lagoon shallows. Thursday Island's reputation of being a maritime badlands also suited his self-image as a dashing adventurer. On one visit he formed a close friendship with the famous Australian flower painter, Ellis Rowan, who sketched flowers while he strode "out on the rocks hunting for corals that he called 'sea blossoms.'"[31]

William wanted his photographs to be both scientific records and "reefscapes" imbued with aesthetic beauties of colour, design and poetry.[32] Skull Reef on the outer Barrier, for example, reminded him – all too poignantly – of a decapitated human head with an "unevaporated tear in its eye."[33] The aesthetic principle of sublimity informed his matching pair of photos labeled "Flotsam" and "Jetsam," depicting the stranded hulk of a mission schooner and some colossal storm-tossed coral boulders.[34]

* * *

At the urging of his homesick wife, William Saville-Kent was forced to leave his position in Queensland after four years to return to England, but in 1893 he published a book, *The Great Barrier Reef of Australia: Its Products and Potentialities*, that encapsulated this magical period of his life on the Reef. A large-size production in super-royal quarto, with 48 photo-mezzotype black and white plates, and 16 hand-coloured chromo-lithographic plates, it was described as an "*edition de luxe*" and sold at the steep price of four guineas.

William's wounded sensibilities and rich talents had combined to produce a masterpiece. Being the first "biography" of the Reef, the book took its many reviewers by storm. He'd sought to show "from an artistic viewpoint" how photographs and coloured chromo-lithographs could combine to capture the stunning visual spectacle of the Reef's coral gardens and marine creatures. While Thomas Huxley and Matthew Arnold were busy arguing over the divided cultures of art and science, William showed how to merge them. *The West Australian* newspaper proclaimed that his book would "make the scientific man an artist and the artist a scientist, and ... inspire the ordinary reader with a desire to be both."[35] *The Scotsman* declared it the most original book on coral reefs since Darwin, and one destined to be "the first authority on the subject." *Nature* claimed that William's photographic methods had added something entirely new to the scientist's methodology: "[his] book contains a series of nature-pictures of the corals such as has never before been submitted to the scientific world."

Reviewers noted that most people in the northern hemisphere would have had no conception until now of the ravishing beauty and riotous colours of a coral reef and its marine inhabitants. "It almost takes our breath away to be suddenly shown one of these plates," wrote the *Cambridge Review*, "we feel we are looking at the thing itself."[36] *The West Australian* thought William's artistry nothing short of genius.

> Unless one has ... seen for oneself the submarine chromatic effects which are more brilliant than the most gorgeous transformation scene conceived, it would be almost difficult to believe that the bright greens, reds, pinks, blues and yellows are the actual colour of forms.[37]

For most if not all these reviewers, he had for the first time revealed the incredible underwater world of coral reefs.

* * *

Now that it is possible for humans to function underwater for long periods of time with their senses intact, and – in the process – to capture vivid, accurate, underwater images of submarine environments and ecologies, we may wonder how marine scientists ever practiced their profession in pre-scuba times. Yet the improvisations of William Saville-Kent, a stiff, non-swimming gentleman scientist of the nineteenth century, enabled him to observe, study and represent sea creatures and their habitats to a remarkable degree without ever actually venturing underwater. Instead of using technologies to enable himself to swim like a fish, William used them to bring the world of the fish to dry land.

Notes

1 Iain McCalman, *The Reef. A Passionate History – The Great Barrier Reef from Captain Cook to Climate Change* (New York: Scientific American/Farrar, Straus and Giroux, 2014), 249–73.

2 Bernard Taylor, *Cruelly Murdered: Constance Kent and the Killing at Road Hill House* (London: Souvenir, 1979), 54–6; Charles Dickens, *The Mystery of Edwin Drood* (London: Folio Society, 1982 [1870]), 51. These details were suggested in a letter, known as "The Sydney Document," sent anonymously by Constance Kent from Australia to the publisher Geoffrey Bles in 1929, and which, though now lost, is reproduced as Appendix II, in Taylor, *Cruelly Murdered*, 373–4; see also Kate Summerscale, *The Suspicions of Mr Whicher, or The Murder at Road Hill House* (London: Bloomsbury, 2009), 296–8; Noeline Kyle, *A Greater Guilt: Constance Emilie Kent and the Road Murder* (Brisbane: Boolarong Press, 2009), 150–3.

3 Anthony J. Harrison, "Museums,"in *Savant of the Australian Seas: William Saville-Kent, 1842-1908*, 2nd edn. (Hobart: Tasmanian Historical Research Association, 1997), 3, http://www.stors.tas.gov.au/au-7-0074-00017.

4 William Saville-Kent to Sir William Flower, 27 January 1891, Sir William Flower, Semi-Official Papers, Australia and New Zealand, National Library of Australia, MS M2843.

5 Charles Darwin, *The Autobiography of Charles Darwin, 1809–1882*, ed. Nora Barlow (New York and London: Norton, 1958), 82.

6 See Adrian Desmond, *Huxley: From Devil's Disciple to Evolution's High Priest* (Reading, MA: Addison-Wesley, 1997) *passim*; Iain McCalman, *Darwin's Armada* (Melbourne: Penguin, 2009), 293–339.

7 William Saville-Kent, *A Manual of the Infusoria: Including a Description of All Known Flagellate, Ciliate and Tentaculiferous Protozoa*, 3 vols. (London: David Bogue, 1880–1).

8 Harrison, "Museums," 7.

9 Harrison, "The Aquarist," in *Savant of the Australian Seas*, n.p.

10 McCalman, *The Reef*, 145–7.

11 Harrison, "Tasmania," in *Savant of the Australian Seas*, n.p.

12 Harrison, "Tasmania," in *Savant of the Australian Seas*, n.p.

13 William Saville-Kent, "Preliminary observations on a natural history collection made on the surveying cruise of HMS Myrmidon," *Papers and Proceedings of the Royal Society of Queensland* 6 (1889): 219.

14 William Saville-Kent, *The Great Barrier Reef of Australia; Its Products and Potentialities* (London: WH Allen & Co., 1893), viii.

15 Saville-Kent, *Great Barrier Reef*, 121–3; Saville-Kent, "Preliminary observations," 225–31.

16 Saville-Kent, "Preliminary observations," 220; see also Art Gallery of NSW, "Collection: Works by Paul Foelsche," https://www.artgallery.nsw.gov.au/collection/works/?artist id=foelsche-paul (accessed 15 November 2017).

17 Regina Ganter, *The Pearl Shellers of Torres Strait: Resource Use, Development and Decline, 1860s–1960s* (Melbourne: Melbourne University Press, 1994), 151–72; J.P.S. Bach, *The Pearling Industry of Australia: An Account of Its Social and Economic Development* (Canberra: Department of Commerce and Agriculture, 1955), 42–60.

18 Harrison, "Travelling North," in Savant of the Australian Seas, n.p.

19 Bach, *Pearling Industry*, 54–5.

20 Saville-Kent, *Great Barrier Reef*, 214 n19.

21 Saville-Kent, *Great Barrier Reef*, 156, 159.

22 Saville-Kent, *Great Barrier Reef*, 13–14, 19, 21, 37.

23 Saville-Kent, *Great Barrier Reef*, 139–40, 180.

24 Saville-Kent, *Great Barrier Reef*, 139–40.

25 Alison Hingston Quiggin, *Haddon The Head Hunter: A Short Sketch of the Life of AC Haddon* (Cambridge: Cambridge University Press, 1942), 80.
26 Saville-Kent, *Great Barrier Reef*, 32–3, 144–5.
27 Saville-Kent to Sir William Flower, 27 January 1891.
28 Saville-Kent, *Great Barrier Reef*, viii.
29 *The Brisbane Courier*, 31 October 1893, 6.
30 Saville-Kent, *Great Barrier Reef*, 40.
31 *The Argus*, 23 August 1896, 6; Harrison, "Travelling North," n.p.; Christine Morton-Evans and Michael Morton-Evans, *The Flower Hunter: The Remarkable Life of Ellis Rowan* (Sydney: Simon & Schuster, 2008), 11, 16.
32 Rosaleen Love, *Reefscape. Reflections on the Great Barrier Reef* (Washington, DC: Joseph Henry Press, 2001), 97–104.
33 Saville-Kent, *Great Barrier Reef*, 25–6.
34 Saville-Kent, *Great Barrier Reef*, 46, 49–50.
35 *The West Australian*, 1 September 1893, 7.
36 William Saville-Kent, "Extracts," in *The Naturalist in Australia* (London: Chapman Hall, 1897).
37 *The West Australian*, 1 September 1893, 7.

9 Frank Hurley and the symbolic underwater

Ann Elias

Introduction

In 1920, when the underwater was still known in the West as a frontier for discovery, well before Jacques-Yves Cousteau colonised the depths with new technologies that enabled divers to breathe and photograph while submerged, the Australian explorer and photographer, Frank Hurley (1885–1962), notified the public through the popular press that he would travel along the Great Barrier Reef, through the Torres Strait to Papua, and descend underwater to film, from below, the wonders of coral reefs.[1] By taking a viewpoint beneath the surface, in a geographical region described by westerners as the "primitive" outer limits of the world, Hurley hoped to excite the public imagination and, in the process, prove his modernity by taking control of the sea's passage to the depths. In his own words, photography was the key to "unlock the portals of the undiscovered World."[2] With a camera he would supply a globalizing market with sensational, spectacular viewpoints on the strange and the unknown. By descending underwater he would take charge of a wilderness that westerners had tried for centuries to master.

However, like William Saville-Kent, whose career as a scientist and photographer of coral reefs Iain McCalman explains in earlier pages, Hurley's fate was to remain bounded by air and land. Despite his desire to seize the underwater frontier, in 1921 he had no practical way of submerging and photographing from below. Instead, enclosed within the island-like boundaries of boats, ships and canoes, Hurley fixated on the ocean's surface. There he experienced the sense of isolation that Pete Hay ascribes to islanders when they consider transcending the encircling sea and are subjected to "an emphatic perceptual boundary; a clearly evident delimiter that is experienced as an edge of primary significance."[3] In Figure 9.1 lies proof of Hurley's sense of isolation from the underwater by the primal "edge" of the surface of the sea, a zone that author and scuba diver, Barry Lopez, describes as "the harrowing transition zone."[4] In Hurley's photograph, it is not the ocean's depths that are sublime but the skin of the sea and the vision of corals lying half-submerged and half-revealed by water, signaling menace, danger, peril and wreck.

As an explorer of the early twentieth century, Frank Hurley frequently traveled long voyages by sea. According to Rachel Carson, in *The Sea Around*

Figure 9.1 Frank Hurley, *Exposed Coral, Torres Strait,* 1921, black and white glass plate negative. National Library of Australia, Canberra, PIC FH/8913 LOC Cold store PIC HURL 299/3.

Us (1951), those who travel on long sea voyages develop a heightened awareness of the sea's horizontal "mantel."[5] But Carson also wrote how the surface of the sea was more than a mantle: it stood for the physical and psychological threshold to the depths. The undersea came to bear a special cultural force which she attributed to the way it remains invisible in plain sight, but palpable as an object of the imagination. The depths, she said, "[fill] the human mind with an even greater sense of awe and mystery."[6] In recent years, however, Margaret Cohen has pointed to the relativity of the term "depths." She notes, for example, how in the nineteenth century "the depth of the "deep" would sink, as new technologies enabled people to push perceptions further below the surface."[7] The significance of the relativity of deepness explains why even the shallow waters surrounding coral reefs at the Great Barrier Reef symbolised, for Frank Hurley, the unknown.

This chapter looks at social circumstances and aesthetic considerations surrounding the photograph in Figure 9.1, especially connections between the image and Frank Hurley's desires, fantasies and memories of the sea and the undersea. It considers the agency of corals and coral reefs, especially their grip on Frank Hurley's imagination as he contemplated the skin of the ocean and the vertical space below. It addresses connections between Hurley's urge to penetrate the tropical underwater and the broader social drive for colonial dominance of islands and islanders in the Torres Strait and Papua region. Above all, this chapter sets out to explain Figure 9.1 as the manifestation of interiority, abstract ideas, knowledge and history, and the intertexuality of representation.

Hurley's expedition to the Torres Strait

A primary goal of Hurley's 1920–1 expedition to the Torres Strait and Papua was to make a film titled *Pearls and Savages* (1921). It was a successful feature-length film, shown worldwide, its popularity aided by a title that, as Stephen Torre observes, exaggerates the dichotomy of "beauty and barbarity," a common colonialist framework in which a contrast is made between the barbaric figure of indigenous peoples and the beauty of their tropical environment.[8]

In 1919 Hurley was under contract to tour Australian public theatres, to deliver dramatized "lecture films" of the Ernest Shackleton Imperial Trans-Antarctic Expedition of 1914–17. He had served as official photographer and had produced a remarkable body of work that showed the terrors of shipwreck on a frozen sea. One day in January 1920, while on lecture circuit he noted in a diary entry how he felt lured by the tropics.[9] By late 1920 he had set out to explore the Australian colonial administered states and territories of the Torres Strait and Papua. His first diary entry was December 1st 1920. There he states that the purpose of the journey was to make films and take photographs on glass plate negatives for a future illustrated lecture of the type that he had just delivered on the Shackleton expedition, except that the new one would entertain audiences with a travelogue of the tropics. He had arranged to visit Anglican missions in the area and make films about their work. In return the missions would assist his journey and aid his travel. Despite having been commissioned to publicise the value of missionary work, Hurley found in the expedition an opportunity to experiment and innovate in locations that were visually seductive.

Coral reefs

By 1900, one of the greatest desires of humankind was to visit a coral reef.[10] The coral reefs that thrive in tropical waters that circle the globe represent an oceanic geography, an underwater zone, that is relatively shallow yet has projected on Western minds intense curiosity, great imagination, but also a keen perception of human frailty. By 1920 the social force of coral reefs was ingrained in Western cultural memory, especially their impact on navigation and the psyches of sailors traversing the surface of the sea, as well as the trouble they had caused for commerce and exploration. But as perspectives shifted downwards towards the vision that lay beneath the waters surrounding tropical reefs, to the flickering apparition of brightly coloured animals and plants viewed from above the water surface in the space below, a different viewpoint attuned to detail and particularity, and realism, began to emerge.

The beauty of a coral reef was compared to "hoar-frost sparkling on the grass, the dew-pearled spider's web spread between the branches of trees, the gem-enthralled butterflies scintillating with sparkling rubies, sapphires, amethysts and emeralds."[11] From the side of a row boat, or from the edge of the water, coral reefs looked like precious jewels. They filled the eyes of the beholder with a desire to get closer. People gazed into coral pools for deeper truths about nature and human existence seeing themselves both distinct from what they observed, and reflected in the vision as if the beauty of the world was made for their aesthetic appreciation and enchantment.[12] But to mariners, who over the ages had experienced the sea viscerally, coral reefs were a liminal presence lurking beneath the surface of the water. And so, the history of Western engagement with coral reef environments is, as Iain McCalman shows about the history of the Great Barrier Reef of Australia, marked by tensions between the sublime and the beautiful.[13]

Especially psychological in Western history are the reefs of the Great Barrier that extend from the Gulf of Papua through the Torres Strait to the coast of Queensland. Before the reefs were properly mapped, foreigners who passed through there imagined huge underwater traps lying in wait, and felt themselves watched by invisible forces. In 1926, for example, when a ship was wrecked on a submerged reef in northern Australia, passengers described the shock and "uncanny sensation" of feeling the vessel's fast forward momentum suddenly immobilised, the ship having been caught in coral as if it had "literally carved for herself a resting place for life."[14] The sensation was like a symbolic death.

Under the water

By the 1920s, the urge to visit coral reefs had developed into a desire to leave the "world of men" and experience a coral reef from under the sea's surface.[15] This was the excitement that captivated Frank Hurley. In the pursuit of a body of work on corals and the underwater, he was proud of the range and potential of his photographic equipment:

> My outfit is of the most complete. Two cinematographs one adapted for underwater. Also two still cameras & a vast quantity of material for every contingency.[16]

But Hurley did not get under the surface of the sea and film with a waterproof cine camera even though a press report at the start of his travels claimed he had "complete diving and underwater apparatus . . . to secure films of secrets of the deep submarine life along the barrier."[17] There is no visual record of Hurley submerged with a camera, or evidence of photographs taken with a camera submerged. In 1921, underwater cinematography and photography were still in their infancy. Undersea images required long exposure times and were only possible in still water for photographers who had plenty of time, and plenty of help. Hurley, though, was constantly on the move in the Torres Strait and Papua as he travelled from island to island recording people and customs and the work of missionaries. Could he swim? There are no diary entries that describe plunging in the sea. There is not even a hint at a passion for swimming, unlike the swimmers of the Romantic era that Robin Jarvis describes in an earlier chapter of this book who thrilled at the sensation of breaking the boundary of the sea's wilderness. Rather, Hurley's diaries suggest his body made very little contact with the water and the underwater. As a medium, as distinct from a concept, a romance, or a mythology of the sea, the underwater suggests itself through the diaries as a challenging prospect. In short, Hurley's diaries infer he was limited to photographing the underwater through air.

Air and surface

It was on the 10th of December 1920, near Cairns on the Queensland coast, during the first adventure to the Great Barrier, that Hurley first sighted a coral reef. At the Endeavour River, he was reminded how Captain James Cook was shipwrecked there in June 1770 when the *Endeavour* hit a submerged mound of coral.[18] As McCalman explains, Cook's decision to sail the reef revealed a distinct environmental misunderstanding and, in the end, he found out the hard way how the life of a reef is mostly submerged. Cook wrote about the unnerving presence of reefs at the Great Barrier, especially the menacing way they tower in the distance above water. He also described them as a "meloncholy prospect"[sic] for the ship's navigation through the "unfathomable" sea.[19]

Once among the reefs and clear water, Hurley felt drawn to photographing the threshold dividing air from sea, a zone of tension because it leads to a more psychological, symbolic space below – a zone readily apparent in Figure 9.1, a photograph of Acropora coral at low tide near Erub (Darnley) Island in the Torres Strait, taken by Hurley in the first months of 1921. Only William Saville-Kent's atmospheric photographs, taken in the late nineteenth century of the surface of sea at the Great Barrier Reef, rival the eeriness of Hurley's image that fixates on the skin of the water. The parts of the image exist in such perfect tension that it is meaningless to guess whether Hurley was attracted to the figure of coral coming forward to meet our eyes, or the lurking ground of the underwater receding from our view. When the National Library of Australia acquired Hurley's collection of glass plate negatives, the cataloguer titled the photograph in Figure 9.1, "Exposed coral." Yet "Submerged coral" seems just as apt, especially for viewers whose perspectives on the ocean are oriented to the underwater. To notice "exposed coral" rather than "submerged coral" illustrates the prevailing cultural tendency to privilege features that relate to land and air rather than sea and the undersea.

Frank Hurley kept meticulous diary notes and drawings. As he sailed the Torres Strait in 1921 he observed and commented on the aura of the reefs. He recorded the moment when:

> We skimmed above a magnificent coral reef. The ocean calm as a vast sheet of liquid glass & we floating on it as if passing through the glass roof of some fantastic garden.[20]

The sea's surface appeared both transparent and indeterminate, a phenomenon created by the meeting of light, air and water that produced a material effect somewhere between solid and liquid – a viscous layer but also a hard unyielding roof. The "roof" of the sea was a common metaphor then. The upper zone of the sea is bounded by a surface or "roof" where water meets air and this is what makes the sea different from terrestrial atmosphere which has no upper boundary, according to James R. Gibson writing in 1979.[21] In Figure 9.1, Hurley tried to capture the sensation of a roof as well as the aesthetic of "liquid glass." He tried to give form to the seemingly ambiguous, irrational, unpredictable materiality of the surface of the sea. The water is shallow in Figure 9.1 but there is anxiety in the image. What was the source?

Wreck

Erub, one of the first islands Hurley visited in the Torres Strait, was then called "Darnley." Hurley was contracted by the *Sun* newspaper in Sydney to write stories about the Torres Strait. He had at his disposal the labour of Torres Strait Islanders who knew the region in all its dimensions including the descending space from land to seabed, and the horizontal expanse across the maze of reefs. As his boat negotiated the submerged reefs, Hurley was glad to have the benefit of traditional knowledge and indigenous navigation. Europeans generally felt this way about indigenous seamen and fishermen. One nineteenth century captain urged that "no one ought to attempt a voyage through the South Sea Islands without carrying an extra crew of this kind."[22] But as "native crews" they represented anonymous labour, and very few were written into history, especially by name.

Suddenly at night during a tropical storm, a crash told Hurley the pearling lugger (sailing ship) he was travelling on had struck a reef.[23] With what he called "sharp coral tongues penetrating the timbers," and the vessel heeling over, the travellers had an anxious wait for the next high tide. Geologists sometimes used the phrase "coral tongue" when describing a specific shape of coral reef formation, and Hurley, who considered his work scientific, was likely to have sourced the term from a technical book. But in the context of the diary, the image of "sharp coral tongues" becomes a distinctly feminized metaphor, inferring the tropical sea was a feminine adversary. To the contemporary reader it serves as a reminder that in the epistemological view where women and "nature" are othered to culture and men, women's bodies are often invoked in accounts of "man's fight with nature."[24] Consequently, when the water was calm Hurley recorded how "the reef girt sea is a placid mirror reflecting islands & clouds."[25] But at other times he recognized that the water was treacherous and the coral, deceptively frail and dainty in a manner that echoed Western perceptions of women.

Memory

As he surveyed the coral reef, it triggered in Hurley a memory of a traumatic time during the ill-fated Shackleton Expedition. In the apparition of coral barbs protruding from the submerged reefs below he discovered an uncanny coincidence of form. He saw a mimicry in miniature of the towering ice formations that had surrounded the expedition ship, the *Endurance*, in 1915. His mind was brought back to the time in the Weddell Sea when the *Endurance* was crushed and sunk by ice. In his diary, Hurley observed how in the Torres Strait "the ocean was littered with outcrops from the submerged reef & looked much like an ice strewn sea."[26]

It was strange to remember the freezing Antarctic while sailing in the hot tropics but the reverie came to Hurley when he looked over the side of the pearling vessel, and noticed the resemblance between coral reefs and icebergs, especially because their mass is mostly submerged. He noted the uncanny way the sight of corals emerging from a glassy ocean surface took him back in time to when the Antarctic sea had been a profound hazard full of deceptive surfaces, and where "lakes amidst the floes [had] the calmness and transparency of pools of glycerine . . . ."[27] For six months, from November 1915 to April 1916, the party was trapped on ice floes. Hurley had experienced the same feeling of precariousness and the "uncanny yet fascinating sensation walking over the bosom of the sea" with Douglas Mawson during the Australasian Antarctic Expedition (1911–14).[28] However, on the Shackleton expedition, Hurley also witnessed two men fall through the ice to the sea beneath, their bodies sucked from air to the underwater with the roof of the sea threatening to close over them like a trap.

To describe the surface of the sea in the Torres Strait as "liquid glass," and in Antarctica as "glycerine," shows that what gripped Hurley's imagination about the ocean was the seduction and deceptiveness of its skin. In general it can be said that the sea was a feminine force for Hurley that left him with a morbid fascination for the membrane that separates what lies above from what lurks below.[29] There in the colonial tropics, in the heat and sun – the antithesis of the frozen Antarctic – Hurley saw an ice-strewn sea and experienced once again the traumatic event of shipwreck that had occurred six years earlier. It was a coincidence that compounded the already intense psychological energy of the coral reefs he surveyed.

Frank Hurley had said his ambition was to "unlock the portals of the undiscovered world."[30] But what beckoned Hurley in the Torres Strait were uncanny sights that he found strangely familiar. Sigmund Freud proposed in 1919 that the uncanny is an aesthetic experience beyond beauty, and often involves frightening feelings arising from memories in the unconscious that recur in the present like an "involuntary repetition."[31] Some people are more susceptible than others. In prongs of corals poking through the mantel of the sea, Hurley recognised Antarctica, although an Antarctica made strange by the comparatively miniature scale of coral reefs substituting as icebergs, and by the materiality of limestone and animal polyps substituting for ice. But the hallucination unleashed anxious sensations, and it is these that resonate in the apparition of corals in Figure 9.1.

The photograph

The main motif of Figure 9.1 is the threshold; the surface that shows simultaneously the above and below and the separation between air and water. Hurley's perspective forces an interaction between the two spaces. For the viewer, this is what gives the image the quality of a dream since the strangeness of being able to view two spaces at once, especially one that is usually hidden by the surface of the sea, recalls the impossible and bizarre visions of a dream. But the surface is also, in part, a space of longing for Hurley because, as indicated, he did not fulfil his goal to photograph while submerged in the sea. Filled with desire, he photographed the thing that most tantalised but also eluded him. Had he succeeded in photographing the underwater, he would have secured a more impressive place in modernity by being remembered as the one who first conquered the frontier of underwater photography in Australia. The surface of the sea in 1921 therefore symbolised for Hurley not only the battle between man and nature but also the struggle between being modern and being constrained by traditional systems of visual representation and mobility. The constraints would only be broken when a new generation, discussed in Margaret Cohen's chapter, "The Shipwreck as Undersea Gothic," deployed new maritime technologies to make the sea's depths reachable.

Hurley took the photograph in Figure 9.1 within days of being shipwrecked in the Torres Strait. The nature it shows is suggestively primitive, savage and unruly. It also captures a long history in which coral is a feared object. The intensity of the image, the way it shows coral writhing underwater but also displays white pinprick gleams of light reflected from exposed spikes, suggest it was a vision that fixed Hurley with a gaze that coral projects onto anxious travellers of the seas. Later in the journey, Hurley would list the kaleidoscopic colours of the reefs – colours abstracted from the picture in Figure 9.1 by the black and white medium – and describe corals as "extravagant beauty."[32] But this phrase, inferring the excesses of nature, also intimates a further anxiety, one anthropocentric in nature and associated with Hurley's uncertainty about the purpose of coral flamboyance. Why do corals look so joyous when so few human eyes will ever see them? Recognising in corals the quality of extravagant beauty brought Hurley to the realisation that coral reefs exist for no special privilege of the human subject. In more recent times this observation has been expanded upon by Alphonso Lingis. While diving in seas around a coral reef, among the "lustrous" shapes and colours of animals and plants, Lingis came to understand that this was "a sphere of resplendent phenomena whose glory, utterly disinterested, calls for no acolyte."[33]

So, rather than experiencing paradise in the legendary exotic tropics, Hurley's encounters in 1921 were confronting on many levels. The threshold, which is so palpable in Hurley's image, lends it a surrealist quality even though the photograph precedes the movement of surrealism by three years. Nevertheless, Susan Sontag's argument throughout *On Photography* is that modern photography exhibits a general, surrealist sensibility and attitude.[34] As in surrealism, Figure 9.1 encapsulates an economical choice of signs that communicate about a psychological transition similar to the exploration of surfaces and thresholds involving water, mirrors, and doorways seen in works by surrealist artists Salvador Dalí, Leonor Fini, René Magritte, Dorothea Tanning, and Max Ernst. Thresholds are emblematic of the functioning of the unconscious as it breaks through the surface into conscious thought, or alternatively of repressed thoughts as they disappear into the depths. As Charles Eldredge shows with Herman Melville's poem "In a Church of Padua" from *Timoleon*, where diving signifies descent into the soul[35], in modern art and literature and in surrealism, depth signified the unconscious, a threshold known as a zone "which separates two opposing worlds ... to cross a threshold is thus to traverse a zone of danger where invisible but real battles are fought out."[36]

Frank Hurley was not a surrealist but in the Torres Strait he was a traveller in a geographical region that Robert McNab names "a territory of great imaginative and cultural value" to the surrealists.[37] When the poet Paul Eluard travelled there in 1924 it was to escape Europe and the memory of the First World War, and to experience the Antipodes, a place conceived of as opposite to Europe and in avant-garde circles that symbolised the primitive unconscious of civilization, thereby promising an alternative to the pretext of the rationality and progress of industrialised society. While Hurley's commercial plan was to film and photograph people, customs, and environments that would result in *Pearls and Savages* and astound Western eyes, there is every reason to suppose that he too was looking for re-enchantment and magic after the inhuman disaster that was the First World War.

"The first and most fundamental feature of the modern idea of nature is a sharp dichotomy between man and nature," wrote J. Baird Callicott in 1992.[38] But what the case of Frank Hurley's photograph in Figure 9.1 shows, especially Hurley's conscious association of the submerged reef and its "sharp coral tongues" with the iceberg strewn sea of the Antarctic, is, as Michael Pollan argues, an example of how the natural world is "cross-wired with human ideas."[39] Hurley's image of submerged corals reminds us, as much as it reminded him, of the precariousness and fragility of our existence. There is a strong sense from the photograph that he explored in a full way the symbolism of the transparency of water, enjoying its ability to reveal what is otherwise concealed as an evocation of psychical processes connected with the mind.

The corals looked back. The gaze of corals – what struck the eyes of Hurley – had a decentering power, a force referred to by Walter Benjamin as the "gaze of nature" and about which he said:

> Wherever a human being, an animal or even an inanimate object thus endowed by the poet lifts up his eyes, it draws him into the distance. The gaze of nature, thus awakened dreams and pulls the poet after its dream.[40]

In 1920, in the "virgin" photographic space of what was erroneously called the "unexplored" regions of tropical coral reefs of Australia, the Torres Strait and New Guinea, Hurley was drawn to an uncanny ecology in which his past took shape again. To adopt Simon Schama's argument about landscape: in the submerged reef in Hurley's photograph, we see a "work of the mind. Its scenery is built up as much from strata of memory as from layers of rock."[41] He endeavoured, through photography, to make the unknown world of the underwater known to people who had never ventured past their own immediate surroundings. But the wilds shaped him, showing him the depths of his own memory. This he combined with cultural knowledge of previous representations of coral reefs and oceans including the stories, myths, and illustrations of James Cook, Jules Verne, perhaps the artists of the Romantic movement and the myriad authors of sea adventure fiction, whom Margaret Cohen discusses in *The Novel and the Sea* (2010) and whose protagonists are often very like Hurley for the way they:

> [b]attle life-threatening storms, reefs, deadly calms, scurvy, shipwreck, barren coasts, sharks, whales, mutinies, warring navies, natives, cannibals, and pirates – in short they have *adventures*, as many such novels emphasized with the wording of their titles. [42]

They are the protagonists of a branch of modern fiction beginning in the eighteenth century that was also the source of imagery for surrealists in the same decade that Hurley explored the Great Barrier Reef. Recollection of multiple visual and textual images combined with Hurley's personal memories of peril and adventure, and his direct experience of the alterity of the coral tropics, to create the intertextual apparition that comes to us through Figure 9.1.

Conclusion

Five years after the Torres Strait photograph in Figure 9.1 was taken, during a return to the region for new work, Hurley described the sea as a "mysterious treasure chest" filled with exotic, rare, and desirable phenomena of nature ready for the taking. He said "there is little to be seen on the surface of the globe these days but the sea-floor opens up limitless avenues to our inquisitiveness."[43] In 1926, with the alterity of the ocean of seemingly less personal concern, he wrote how the ocean's future value lay in entertainment and the possibilities that it could bring to the pleasure of humans. It was an attitude that foreshadowed the growing tendency to conceive of the sea as a limitless human resource. Hurley's involvement in the mass entertainment industry, the part he played in tourism and exploration, also shaped his interest in natural history. In the materiality of natural history photographs, that he travelled the world to secure for mass distribution, was proof that every corner of the globe would soon be represented, known and conquered. In the 1950s a new generation of photographers, explorers, entertainers and consumers inherited Hurley's sentiments, and in that decade, the underwater was opened up to human inquisitiveness in an unprecedented way.

We know that the reading of images changes with time. Twenty-first century viewers are likely to notice how the black and white medium used by Hurley, has the sad consequence of abstracting the features of the living coral animal from the image, bleaching the image of colour and creating instead an effect of a maze of white skeletal forms. Suggested by the medium of black and white is a lack of corporeal life. This is a melancholy reading of the image but it captures the image of coral we fear today: not a lurking, hidden menace for travellers, but a bleached remnant of what is still, precariously, a vital part of the world's ecosystem.

Notes

1 Frank Hurley, "A Great Enterprise: Filming the Great Barrier Reef," *The Richmond River Herald and Northern Districts Advertiser* (New South Wales), 7 December 1920, 1.
2 Frank Hurley, *Argonauts of the South* (1925) cited in John Thompson, *Hurley's Australia: Myth, Dream, Reality* (Canberra: National Library of Australia, 1999), 2.
3 Pete Hay, "What the Sea Portends: A Reconsideration of Contested Island Tropes," *Island Studies Journal* 8, no. 2 (2013): 216.
4 Barry Lopez, "Searching for depth in Bonaire" (1999) cited in Hay, "What the Sea Portends," 221.
5 Rachel L. Carson, *The Sea Around Us* (London: Penguin Books, 1956), 23.
6 Carson, *The Sea Around Us*, 148.
7 Margaret Cohen, "Denotation in Alien Environments: The Underwater *Je Ne Sais Quoi," "Representations"* 125, no. 1 (2014): 105.
8 Stephen Torre, "Tropical Island Imaginary," *Etropic: Tropics of the Imagination 2013 Proceedings* 12, no. 2 (2013): 251.
9 Frank Hurley, "Frank Hurley's diary, when on tour with the Shackleton Picture, 21 November 1919–20 January 1920," National Library of Australia, MS 888, Papers of Frank Hurley 1912–1962, Series 1, item 6, 63.
10 For example, in 1876 Reverend Tennison Wood wrote in *The Australasian* newspaper that "for my own part, I must say that to see a coral reef had been one of the great desires of my life". See J. E. Tenison-Woods, "A Trip to a Coral Reef," *The Australasian*, 18 October 1879, 6.
11 Eucalyptus, "Farm & Garden," *Townsville Daily Bulletin* (Queensland), 26 January 1932, 7.
12 Jane Bennett discusses enchantment and its "agentic capacities" in Jane Bennett, *Vibrant Matter: A Political Ecology of Things* (Durham, NC: Duke University Press, 2010), xvi.
13 Iain McCalman organises the chapters of his book into "Terror," "Nurture" and "Wonder." See Iain McCalman, *The Reef: A Passionate History* (Melbourne: Viking, 2013).
14 The ship that ran aground on coral was the *Cooma*. See "Circumstances of the Grounding: 'An Uncanny Sensation'," *Northern Herald* (Queensland), 14 July 1926, 38.
15 For example, concerning underwater photography in the Bahamas see "Under-Sea Photography," *The Argus* (Melbourne), 18 February 1916, 5.
16 Frank Hurley, "Being an Account of a Voyage to the Barrier Reef. The Torres Straits Islands and New Guinea. From 2 December 1920. Islands Visited, Darnley, Murray, Mabuiag, Moa," Papers of Frank Hurley, 1912–1962, National Library of Australia, MS 883 Series 1, item 7, 1–2.
17 Frank Hurley, "A Great Enterprise," 1.

18 Hurley, "Being an account of a Voyage to the Barrier Reef," Series 1, item 7, *op.cit.*, 14.
19 James Cook, *The Journals of Captain James Cook: The Voyage of the Endeavour, 1768–1771*, ed. J. C. Beaglehole (Cambridge: Cambridge University Press for the Hakluyt Society, 1968), 361.
20 Hurley, "Being an Account of a Voyage to the Barrier Reef," Series 1, item 7, *op.cit.*, 93–4.
21 James J. Gibson, *The Ecological Approach to Visual Perception* (Boston: Houghton Mifflin, 1979), 21.
22 David A. Chappell, *Double Ghosts: Oceanian Voyagers on Euroamerican ships* (New York and London: M.E. Sharpe, 1997), xvi.
23 Hurley, "Being an Account of a Voyage to the Barrier Reef," Series 1, item 7, *op.cit.*, 22.
24 "Man's fight with nature" was a phrase commonly applied from 1900 to 1940 to the fear of things including insects, beasts, the desert, the sea, and disease. See, for example, "Man's Fight with Nature: Professor Ray Lankester's New Book, 'The Kingdom of Man'," *The Daily News* (Perth), 23 March 1907, 13.
25 Hurley, "Being an Account of a Voyage to the Barrier Reef," Series 1, item 7, *op.cit.*, 14.
26 Hurley, "Being an Account of a Voyage to the Barrier Reef," Series 1, item 7, *op.cit.*, 24.
27 Frank Hurley, "The Imperial Trans-Antarctic Expedition Diary," 16 December 1914, in *The Diaries of Frank Hurley, 1912–1914*, eds. Robert Dixon and Christopher Lee (London and New York: Anthem Press, 2011), 13.
28 Frank Hurley, "Where Icy Blizzards Blow," *Sun*, 26 June 1927, 22.
29 Feminine metaphors of the sea abound in nineteenth century literature, including Jules Verne. Captain Nemo to Professor Aronnax: "Ah, monsieur, to live in the bosom of the sea! Only there can independence be found! There I recognize no master! There I am free!" Jules Verne from *Twenty Thousand Leagues under the Sea* (1870) quoted in David Lambert, Luciana Martins and Miles Ogborn, "Currents, Visions and Voyages: Historical Geographies of the Sea," *The Journal of Historical Geography* 32 (2006): 480.
30 Hurley cited in John Thompson, *Hurley's Australia*, *op.cit.*, 2.
31 Sigmund Freud, "The Uncanny" (1919), in *The Standard Edition of the Complete Psychological Works of Sigmund Freud*, ed. and trans. John Strachey vol. 17 (London: Hogarth Press, 1964), 247.
32 Frank Hurley, *Pearls and Savages: Adventures in the Air, on Land and Sea – in New Guinea* (New York and London: Putnam, 1925), 82.
33 Alphonso Lingis, "The Rapture of the Deep," in *Excesses: Eros and Culture* (Albany, NY: State University Press of New York, 1983), 9.
34 Susan Sontag, *On Photography* (Auckland: Penguin, 1982), 54–5.
35 Charles C. Eldredge, "Wet Paint: Herman Melville, Elihu Bedder and Artists Undersea," *American Art* 11, no. 2 (1997): 110.
36 Emma Cocker quotes Marcel Griaule's description of a threshold from George Bataille's *Critical Dictionary* in Emma Cocker, "Desiring to be Led Astray," *Papers of Surrealism* 6 (2007): 11.
37 Robert McNab, *Ghost Ships: A Surrealist Love Triangle* (New Haven and London: Yale University Press, 2004), 69.
38 J. Baird Callicott, "La Nature est morte, vive la nature!," *The Hastings Center Report* 22, no. 5 (1992): 16.
39 Michael Pollan, *The Botany of Desire: A Plant's-Eye View of the World* (New York: Random House, 2002), 70.

40 Walter Benjamin from "On Some Motifs in Baudelaire" quoted in Willem van Reijen, "Breathing the Aura – The Holy, the Sober Breath," trans. Claudia Kotte, *Theory, Culture & Society* 18, no. 6 (2001): 43.
41 Simon Schama, *Landscape & Memory* (London: Fontana Press, 1996), 6–7.
42 Margaret Cohen, *The Novel and the Sea* (Princeton: Princeton University Press, 2010), 3.
43 Frank Hurley, "The Floor of the Coral Sea: Among the Blue-green Opal," *Sun* (New South Wales), 7 February 1926, 17.

10 The shipwreck of reason

The Surrealist diver and modern maritime salvage

Sean Theodora O'Hanlan

In an episode in *Soluble Fish* (1924), André Breton and his Surrealist companions outfit themselves in glass diving suits in order to convene in an "underwater command post," as he writes:

> We seldom enter this room unless we are wearing glass diving suits that permit us, as we go where the tipping planks take us, to meet when necessary at the bottom of the sea...We are the creators of wrecks; there is nothing in our minds that anyone will manage to set afloat again.[1]

As birds commingle with fish beneath the waves, this inverted palace, complete with an underwater smoking room, evokes a Rimbaudian *salon au fond du lac.* The Surrealist divers dwell within a maze of interconnected interiors, including a vaporous bedroom and, in something of a modern refurbishment of Plato's allegory of the cave, a cleverly designed underwater theater that projects "the shadow of ideas." As they move freely in this *autre monde*, the Surrealists undertake the "great immaterial destruction" of outmoded and functional machines. Disavowing rational principles of modern technological progress, they divert these devices from use:

> For the moment we are taking the machines that have ceased to be useful, and also a few others that were beginning to be useful, to the bottom of the water, at great expense, and it is a pleasure to see the mud voluptuously paralyze things that worked so well.

In this episode, Breton introduced two interrelated, recurrent tropes of the shipwreck and diver. In his paradoxical description "creators of wrecks," Breton explicitly described the Surrealist project in terms of underwater submersion and salvage.

The present study takes as a point of departure this configuration of the diver and shipwreck, as they constitute an aquatic current within Surrealist art and literature of the interwar period. Through the diver-shipwreck configuration, Breton carried out contemporary critiques of the individual and collective unconscious, linking the Surrealist exploration of the depths of the mind to the material conditions of the historical moment. In so doing, Breton limned present destruction and historical salvage. The Surrealists both created – and came to explore – the wreck.

* * *

There are many routes the present study could take in charting Breton's undersea imaginary.[2] Far from exhaustive, what follows is but a sounding, intended to highlight the depth of such an engagement through a focus on Breton's writings and a selection of related texts and images created between 1919 and 1939. Participating in a long tradition of subversive quests at sea, Breton frequently presented the Surrealist project as an intentionally anachronistic, fated collective voyage across vast distances and into great depths. Departing from Enlightenment concepts of vision, selfhood, and representation, the Surrealist quest possessed none of the pragmatic imagination of actual maritime craft or undersea salvage. Excessive in its projection of a boundless, full-fathom elsewhere, the journey was resolutely non-instrumentalized, undertaken with broken instruments (the compass "constrained to south") and at great risk to the subject. Perpetually oriented beyond the limits assigned by bourgeois ethics and aesthetics, Breton declared Surrealism to be "a function of our dépaysement from everything."[3] *Dépaysement*, which literally translates to a state of *un-landing*, is a term for the physical and psychological experience of displacement, disorientation, or exile. The term was central to the Surrealist defamilarization of the quotidian. Presupposing literary precedents as well as world-historical events, the diver and shipwreck served as metaphors for the unmooring of the rational subject, as well as a collective departure from the existing world order.[4]

"The Surrealists," wrote Allan Sekula in *Fish Story*, "were the last aesthetic movement to claim the sea with any seriousness."[5] In the paradoxical erasure of the sea in the age of global capital, Sekula argued that a historical and conceptual shift in the modern subject's ocean imaginary resulted in the "death of seascape." According to Sekula, the Surrealists both revived the romantic sublime and attempted to "domesticate" it; a contradictory engagement, he warned, in which Surrealist representations of the sea risked resembling commercialized mass culture. Though Sekula expertly situated Surrealism with regard to such precedents as Marcel Proust, *Fish Story* does not address Surrealist sources, figures, or works of art in any specificity. Moreover, Sekula's analysis remains at the sea's surface, as an outsized focus on proximity and distance largely fails to account for the Surrealists' equaled – and perhaps more generative – engagement with the depths of the sea. Reconceived here, it was precisely this dialectic of surface and depth that Breton and his companions recurrently sounded as an image suspended between the individual and collective unconscious. At once living and dead, the psychologized sea was both a space of potentiality governed by the imagination and a site of lost possibility. If it was a mirror that reflected the Surrealist subject, its depths contained worlds. The result is something like the doubled vision naturally effected by a *fata morgana* in certain atmospheric conditions, in which the ship upon the horizon appears doubled and inverted: the spectator perceives the ship and its wreckage at once.

The Surrealists worked against the mapping of rational Enlightenment aesthetics onto the realm of experience and perception, instead insisting upon the power of the imagination and the subjectivity of vision.[6] If the group borrowed much from the sublime, the Surrealist dive into the wreck was also inherently modern, unique to the conditions of the age. Breton often insisted that Surrealism could only be understood in historical relation to the psychological and ethical shocks of the First World War. Given the unprecedented carnage inflicted by newly developed militarized submarine technologies during the conflict, the shipwreck and diver were emblems of the collective trauma of modern sea power. In the wake of WWI, writers, politicians, and historians across Europe employed imagery of deluge and shipwreck in order to explicate how the redrawn world map of 1919 correlated to profound, if less visible, consequences.[7] Writing with an atlas in hand, the poet Paul Valéry described how the "shipwreck" of the preexisting global system correlated to a crisis of individual and collective consciousness:

> Through the obscure depths of history we could make out the phantoms of great ships laden with riches and intellect gone down with all their men and all their machines into the unexplorable depths of the centuries ... the total ruin of those worlds had as little significance for us as their very existence ... And we see now that the abyss of history is deep enough to hold us all.[8]

In order to image the psychological crisis that accompanied the political remaking of the world order, Valéry evoked a submarine space in which broken "ships" of modern nation states (France, England) and those of bygone civilizations (Nineveh, Babylon) lay on the seabed beside the *Lusitania*, the shipwreck of recent memory. Bridging psychological, political, and historical registers, metaphorical and real historical events of shipwreck converge indiscriminately, "beautiful vague names" in an eerie landscape.

These coexistent registers within Valéry's cultural critique were foundational to the young Breton and his compatriots as he situated Surrealism's embarkation somewhere between the dictum "après nous, le Déluge" and Rimbaud's "Après le déluge." Max Ernst formulated an origin story for the movement, casting the group as navigators of an ark that emerged from the Dada "flood":

> Surrealism was born in the middle of the deluge of Dada, when the ark struck a ridge/peak. The explorers did not have the slightest desire to repair their boat, or to settle upon the island. They preferred to dive in.[9]

The soluble subject: between perception and cognition

On July 4, 1925, Breton saw J.E. Williamson's documentary film, *Les Merveilles de la mer*, and described it in a letter to his wife Simone Kahn: "Last night I saw *The Wonders of the Sea* ... There are terrible parts, but the walk of the divers against the currents—that's ballet. And then the octopus is so beautiful."[10] Less than two weeks later, Breton published the first installment of "Surrealism and Painting" in the fourth issue of the journal *La Révolution Surréaliste*. The text began with Breton's famous declaration:

> The eye exists in a savage state. The Marvels of the earth a hundred feet high, the Marvels of the sea a hundred feet deep, have as their sole witness the wild eye ... But who is to draw up this scale of vision?[11]

Breton's reference to the marvels of the sea was not incidental, but was in fact the popular title given to the Williamson film when it was shown at the Vieux-Columbier theater in Paris that summer. As demonstrated by his collage-like incorporation of the film's title, Breton found an analogue to the Surrealist reorientation of visuality, experience, and perception in the exploration of the marvels of the sea accessed through underwater film.

That Breton saw Williamson's film in 1925 attested to the rapid international circulation of the undersea environment in several forms within popular culture at the very moment a "new scale of vision" was also pursued through advancing technologies of underwater film. The Williamson "picturization" of Vernian depths in 1917 was compared to the technological innovations of the railway, radio, and airplane. As a result of the "marvelous" technological innovations of undersea cinema, one critic explained that moviegoers "in hundreds of moving picture theaters scattered over many lands ... [are] transported bodily to the bottom of the ocean."[12] However, praise for the modern "age of wonders" was accompanied by a warning: "It is also the age and day of blunted vision and warped perspective." As the modern subject attempts to "grasp, analyze, and mentally catalogue" these advancements, they risk sensory derangement as "the ear rings, the eye swims, and the brain becomes confused."

To lose one's sense of scale and value: It was precisely this disorientation of habitual perception and salvage of attention that Breton sought as he referred to the physical nature of vision underwater in explicating Surrealism's perceptual experience in unfettered thought. In conceiving of an aesthetic predicated upon experiential sensation rather than mimetic representation in "Surrealism and Painting," Breton asserted the need to create an unhindered, "new scale of vision" beyond the visual language of the real. This new, more "penetrating" sight was doubled, attuned "to that which is not visible" in addition to "conventional exchange of signals."[13] The "marvels of the sea" were not only metaphors for this new "scale" of vision, but a literal reference to both the limitations and radical possibility of "penetrating sight" in a *dépaysé* environment.

In order to fully attend to Breton's conception of Surrealist visuality, it is helpful to consider how the diver-shipwreck configuration subtended Breton's earliest articulation of the group's exploration of the individual unconscious, which were often represented in the form of ocean depths. In the five years preceding Breton's definition of Surrealism's "pure psychic automatism" in the *Manifesto of Surrealism*, the tropes of the diver and the shipwreck were integral to the nascent Surrealist group's challenge to the rational subject through automatic expression. Despite the importance of automatic writing, automatism was always inherently visual, as the poet Robert Desnos represented the Surrealist exploration of the unconscious in enigmatic seascapes filled with maritime motifs.[14] In one such automatic drawing, the word "*ou*?" or "where?" in the center of the sky (with several divergent arrows) suggests that the expedition is without destination, as a small skiff threatens to collide with a large ocean liner, christened "*or*," or "gold" (Figure 10.1). Desnos, who possessed a "romantic taste for shipwreck," intimates the fatal depths below in the form of a mysterious, pale hand that emerges from the sea's surface in the foreground.

From the very outset of the group's collaborative work between 1919 and 1923, the Surrealists compared the profoundly altered, reoriented "vision" in automatism or dream to the deep-sea diver's perception underwater. Equipped with a "diving bell of automatism," Breton described automatic practices – verbal, visual, and written automatic practices, trance

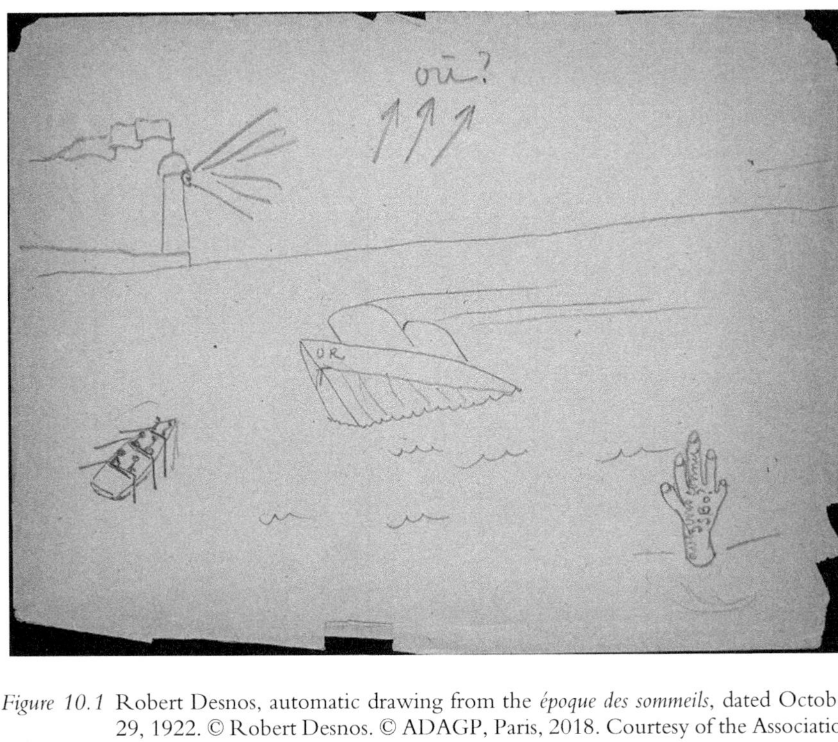

Figure 10.1 Robert Desnos, automatic drawing from the *époque des sommeils*, dated October 29, 1922. © Robert Desnos. © ADAGP, Paris, 2018. Courtesy of the Association Atelier André Breton. www.andrebreton.fr.

states, hypnosis, and free association – as instruments that enabled the subject to access the "never seen." Already in 1919, Desnos rendered the perspective of the dreaming subject in a small ink drawing on the final page of "Ode to Coco" (Figure 10.2). With a truncated hull and foreshortened smokestacks, Desnos depicts the ocean liner and its subtle wake from below. As a reader, we are beside Desnos beneath the waves in dream; looking up, we assume the view of a "fish of dream" described in his text at the top of the page.

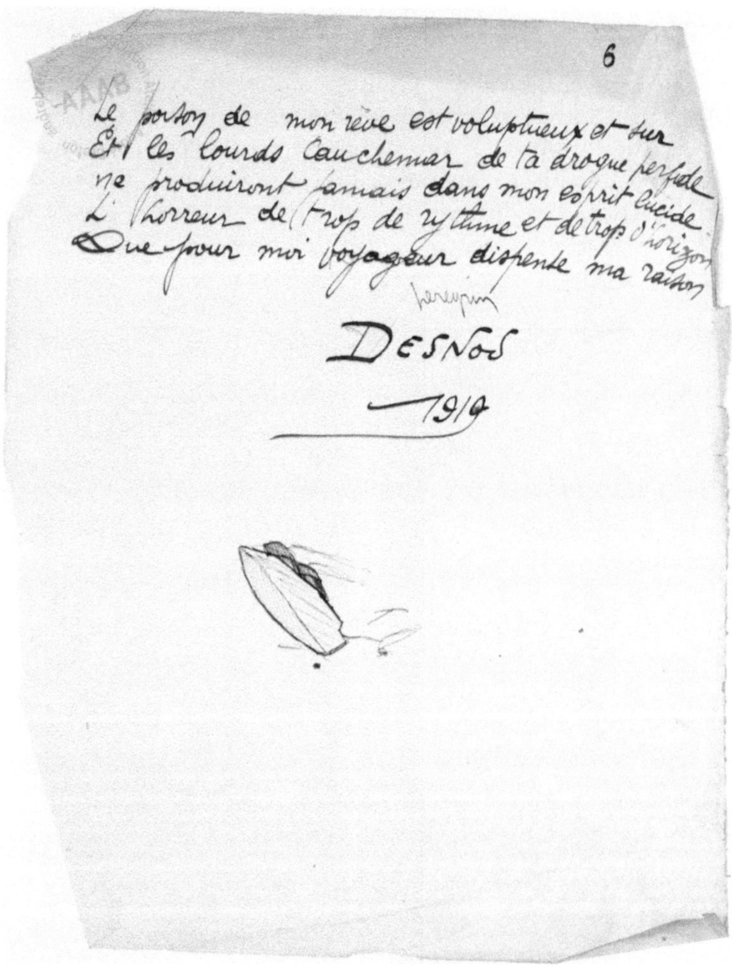

Figure 10.2 Robert Desnos, detail in "Ode to Coco" manuscript, 1919. © Robert Desnos. © ADAGP, Paris, 2018. Courtesy of Association Atelier André Breton. www.andrebreton.fr.

Desnos' image prefigured the central image of Breton's *Soluble Fish* (1924). Breton originally conceived the *Manifesto of Surrealism* as a brief introduction to the thirty-two episode automatic text; composed simultaneously, it constitutes something like the unconscious depths of Surrealism's foundational document. Indeed, in her analysis of the *Manifesto* manuscript, Marguerite Bonnet has noted that Breton edited his well-known directive for the techniques of automatic writing, the final form of which reads, "Write quickly, without a preconceived subject ... the first phrase will come spontaneously, as it is true that with every second there is *a phrase unknown to our consciousness...*" Breton had originally formulated the sentence in keeping with the aquatic imagery of *Soluble Fish*: "*a phrase that has passed imperceptibly into the depths of the water.*"[15] This latent reference – as it reveals the interrelation of the two works, "poisson soluble" in the original "poisson invisible" – indicated that the "soluble fish" was, for Breton, intended as an emblem of automatic writing itself (a fact supported by the text's very title.)[16]

Georges Malkine seems to have illustrated the image as precisely that – an emblem – in 1925. As the group established operations at the Bureau of Surrealist Research, Breton asked Malkine – the sole artist among the signatories of the *First Manifesto of Surrealism* – to design a Surrealist "logo" for internal use at headquarters. An examination of one of the resulting drawings, *Sirens*, underscores the centrality of aquatic imaginary to early Surrealist interrogation of the subject. In the drawing, the disembodied head of a mysterious woman gazes into an aquarium. Malkine registers the slight optical distortion where her chin meets the level of the water in the tank, the top frame of which cuts across her face and obfuscates her eyes. The eyes reappear, displaced and multiplied in the form of seven transparent "fish-eyes" with ciliated lashes for fins. In the form of the soluble fish, Malkine depicted the submerged sight in "soluble thought," integrated in the viewpoint of the subject – "I" and "eye." The subject both peers inward, at the depths of the mind, and returns a gaze *from* within, becoming simultaneously subject and object of a gaze (Figure 10.3). In 1926, Malkine employed a vivid palette to create a painting based on the drawing, integrating the visual effects of underwater optics in the subtle gradation of water in depths (Figure 10.4, see plate section).

The perils of the Surrealist collective

While this set of submarine references might seem unique to the operations of psychic automatism, the diver and the shipwreck were redeployed to historicize the movement's interventions with the publication of the *Second Manifesto of Surrealism* in 1929. As Breton reflected upon Surrealist interventions in the five-year period since the *Manifesto* (1924-29), these tropes served as vehicles through which he strategically worked to assimilate the group's experimental aims to the goals of collective action, as they were defined, in large part, by the French Communist Party (PCF). Under harsh scrutiny from the intellectual left, Breton continued to assert that the primary aim of Surrealist research was

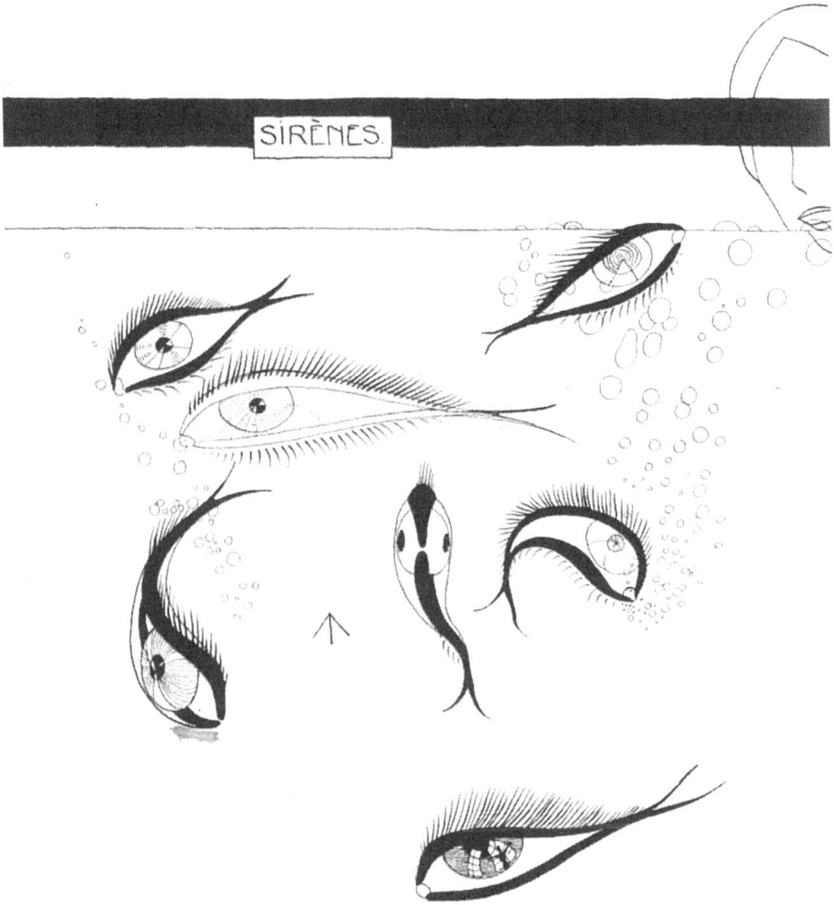

Figure 10.3 George Malkine, *Sirènes*, 1925, India ink on paper. © ADAGP, Paris, 2018. Courtesy of the Association Atelier André Breton. www.andrebreton.fr.

to access an as-yet-undiscovered point that mediated dialectical oppositions, framing Surrealism as a cartographic endeavor. As he faced criticism from within and without, the ethical ambiguity, aspirations, and risks of maritime craft in many ways suited Breton's uncertainty as he worked to concretize the group's collective ambitions. In the process, his use of these tropes underscored the inherent challenges – if not limitations – of his attempt to bridge Freudian thought and dialectical materialism.

In the *Second Manifesto*, Breton provided a historical overview of the movement to date that reaffirmed the centrality of automatic expression, but also reframed Surrealism's group identity in materialist terms. In setting out these aims, Breton made a historical distinction between Surrealism and its predecessor, Dada. While Dada, Breton writes, was interested in "the torpedoing of the idea in the midst of the sentence which is articulating it," Surrealism "through its appeal to automatism, was involved in sheltering from this torpedoing a building of some sort: something like a Flying Dutchman (*vaisseau fantôme*)."[17] Here, Breton seemed to elaborate upon the image of "wrecks" in *Soluble Fish*, defining the movement as ambiguously "constructive and destructive," in seeking out a *point idéal* in which oppositions of life and death cease to exist. In what at first appears to be a historical and conceptual distinction, Breton seemed to differentiate Surrealism's project of "preservation" from the explicitly destructive mission of Dada. However, the distinction is historically ambiguous. With the image of the torpedo – a modern technology invented in the nineteenth century and associated with militarized maritime technology of WWI – Breton located Dada firmly in the present age. Originating in the seventeenth century, the romantic trope of the Flying Dutchman is not only anachronistic, but also a harbinger of death to sailors who encounter it. The ghost ship's function, then, presented a profoundly paradoxical image of "preservation," performing a certain "sheltering" only insofar as it endures across time. Situating Surrealism before and after Dada's "torpedoing" in the present, Breton denies a clear distinction, as if to assert that Surrealism passed *through* Dada's spirit of destruction.

On the eve of 1930, the anachronistic, staid image of the Flying Dutchman also reflected the perils of the group's collective ethos. As several former Surrealists responded to the *Second Manifesto* by publishing a scathing critique of Breton entitled "Un cadavre" (1930), modeled after Breton's own pamphlet of the same name six years earlier, Jacques Prevert seized upon the reference to the Flying Dutchman in order to mock Breton's leadership role, sardonically suggesting that Breton, upon "seeing the Flying Dutchman pass in dream," fancied himself a naval officer in the mode of the bumbling villain Captain Bordure in Alfred Jarry's *Ubu roi*.[18] Dispelling the pathos Breton attached to the image, Prevert asserted that it served as an emblem of Breton's moralized, aestheticized retreat from the real, concluding that Surrealism failed to engender collective action, and therefore lacked any true liberatory import. Breton later remarked of the period of agitation after the *Second Manifesto*: "Surrealism at that time reminds me of a superb and demasted ship, which could from one minute to the next either have gone to the bottom or have triumphantly reached the land of which Rimbaud spoke where we would at last know 'the true life.'"[19]

Historicizing Surrealism: the politics of a collective unconscious[20]

Tethering the individual to a collective unconscious, the shipwreck and diver came to serve as dialectical images of historical salvage by the mid-1930s. As the Surrealists endeavored to communicate their revolutionary ethos beyond France's borders through exhibitions, publications, and lectures, Breton's articulation of the interrelation of aesthetics and politics became increasingly urgent with the rise of fascist regimes across Europe. The political application of Surrealism's historiographic project, expressed through Surrealist salvage, becomes apparent in a constellation of texts and images published in 1935. In his "Speech to the Congress of Revolutionary Writers and Artists" in June 1935, Breton asserted that the Surrealist artist had a duty to "inventory" and "shelter" poetic thought as it formed a collective unconscious:

> This property, possessed here and there by certain artistic works, can appear to us only as a function of their very particular situation in time, of that air of being figureheads at the prow of a ship that they assume in relation to the historical circumstances that unleashed them.

For Breton, particular artists and writers like Charles Baudelaire function as "figureheads at the prow of a ship" insofar as they sail forth from their specific historical circumstance and into the present. In so doing, they "bring about a perfect balance between the inner and the outer," finding "authentic" equilibrium in their individual creative production even as they negotiate historically situated conditions of social upheaval.[21]

Breton's image becomes all the more clear if we consider "In the Cemetery of Ancient Ships [Galleys]", a suite of photographs by Bill Brandt depicting salvaged figureheads that appeared in the journal *Minotaure* just months earlier (Figure 10.5).[22] Brandt took the uncanny images in the Valhalla gardens on the low-lying Isles of Scilly, located 28 miles off the coast of Cornwall, England. Ships so frequently wrecked upon Scilly's treacherous shallow reefs that the collection of salvaged figureheads at Valhalla was established in the 1830s, named after the mythological Norse palace of the dead to honor lost sailors.[23]

In Brandt's images, the denizens of the garden converged in uncanny juxtapositions of time and place. By providing accurate identifications, Brandt's photographs were essentially archeological in nature. On the top right, for example, the merry tonsured figurehead that once adorned *The Friar Tuck* – a merchant clipper ship that ran aground in a hurricane while returning from China to Liverpool with a freight of tea in 1863 – strains forward to meet the elegant "Prima Donna" that sat at the prow of the *River Lune*, an iron barque that struck the rocky coast in a thick fog in 1879.[24] The final image captures a disintegrating hull of an unidentified ship on Scilly's shore. This wreckage on the beach functions as something of a *memento mori*, setting out a dialectic of destruction and preservation. By linking the salvaged figureheads to the destroyed vessel, this configuration underscores the displacement of these objects from their geographical and temporal origin, as well as from their historical function.

Brandt's photographs were published on the page that immediately followed Dr. Pierre Mabille's important article, "In Praise of Popular Prejudices," in which the physician and anthropologist illustrated his theory of a physical and psychological "double unconscious" according to a topographical schema. In his formulation, Mabille situated the individual's "visceral unconscious" in relation to a collective "unconscious of oblivion" in which external elements of social life

Figure 10.5 Bill Brandt, "Au cimetière des anciennes galères. Scilly Isles, Angleterre," in *Minotaure* 6 (Winter 1935): 4. Arno Press. Reprint by permission of Éditions d'Art Albert Skira. DACS.

circulate and transform. Summarizing this "geologically rich and unfathomably deep" geography of stratified intellectual and social efforts, Mabille writes:

> A vast submarine reservoir which all the cultures, all the studies, all the reasoning of minds and will, all the social revolts and struggles undertaken are brought together in a formless vessel in which the elements are digested, rot, and mingle as they disintegrate.[25]

As the "tides" of the individual unconscious flow within the mire of human history, social conflicts, revolution and war are capable of activating this great common capital from time to time, surging up from the latent depths to act upon the surface.

If Brandt's photographs of Scilly served in part to illustrate the theoretical conceit of Mabille's vast submarine fund, they also contravened dialectically in Breton's articulation of Surrealism's historiographic project at large. In a second text published the same year, *The Political Position of Surrealism*, Breton paraphrased Mabille as he stated that certain works of art, suffused with the unconscious agitation of their historical moment, withstand time and flash up in legibility as a "collective treasure." Thus, Breton articulated his most extended statements on Surrealism's political position, the function of the artist, and the historical interrelation of aesthetics and politics through the descent and ascent into a collective unconscious.

Conclusion: Surrealist salvage spectacle, 1936–2017

On July 1, 1936, Salvador Dalí delivered an infamous lecture from inside a diving suit at International Surrealist Exhibition in London. A photograph taken in the central gallery reveals Dalí in a canvas divesuit and a copper and brass diving helmet adorned with coral and a wine glass. He appears calm as he sits in the company of Surrealist compatriots E.L.T. Mesens, Paul and Nusch Eluard, and Diana and Rupert Lee (Figure 10.6). In his autobiography, Dalí recalled that a "serious" engineer "from the diving-suit establishment" was present to bolt his helmet into place. The expert evidently left, as Dalí, "equipped to descend into the subconscious," nearly suffocated less than ten minutes into his inaudible speech. After he received assistance removing his helmet, the audience applauded when he returned to the podium, assuming that the rescue had been planned.

If Engels was correct in his assertion that history occurs twice, then surely Dalí's entry to the London Exhibition in a diving suit, carrying a sabre and accompanied by two Irish wolfhounds, was the tragic image's recurrence as farce.[26] As a result of Dalí's stunt, the deep-sea diver was transformed into an icon of the Surrealist project within the popular imagination. As Dalí's stunt circulated as an amusing anecdote conveying the group's penchant for the fashionably outlandish and bizarre, the Surrealists' concerted efforts to achieve the internationalization of the movement in the name of non-national, collective work of interpretation was reduced to spectacle. The episode was noted widely in the international press, and

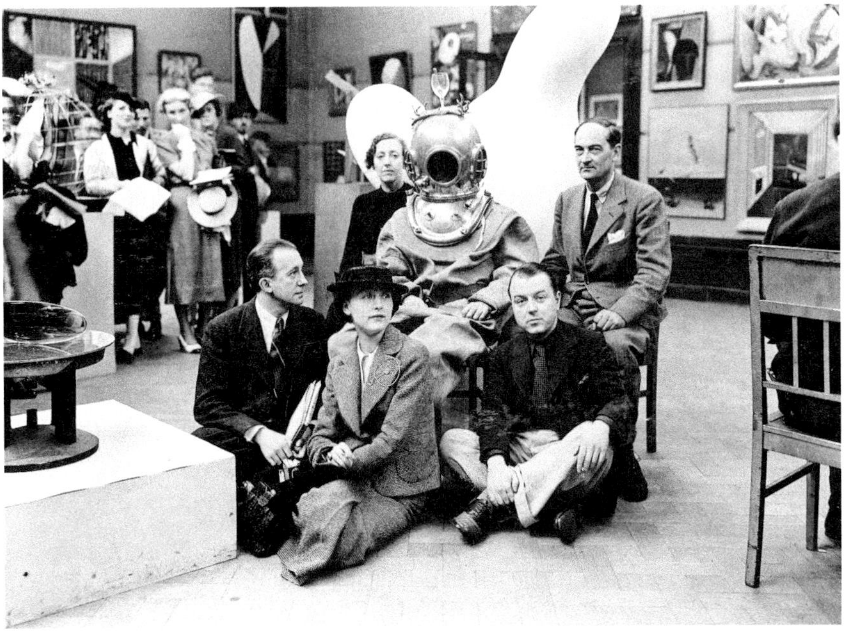

Figure 10.6 Photographer unknown, Salvador Dalí (in diving suit), pictured with Diana Brinton-Lee, Rupert Lee, Paul Eluard, Nusch Eluard, and ELT Mesens at the *International Surrealist Exhibition* in London in June 1936. © Tate, London 2018.

parodied in a cartoon published in *The New Yorker* in the winter of 1937, captioned "A Surrealist Family has the Neighbors to Tea." As their host sips his tea in a diving suit, the family's two shell-shocked guests huddle awkwardly around a cauldron in a sitting room decorated with recognizable Surrealist accouterments (Figure 10.7). In this moment, the Surrealist diver that had pervaded the Surrealist imaginary from its very beginnings entered the dangerous space of the commercial gallery; in many ways, the figure has never surfaced.

As proof, we need look no further than Damien Hirst's "Treasures from the Wreck of the *Unbelievable*" exhibition, held concurrently at the Palazzo Grassi and Punta della Dogana in Venice in 2017.[27] The show displayed a collection of marvels salvaged from an imaginary 2nd-century shipwreck, the *Unbelievable*, purportedly recovered in the Indian Ocean after more than two thousand years.[28] The entirely fictional story of the collection corresponded to an elaborate undersea salvage operation. In addition to the malachite, marble, and gold treasures in each room, extraordinary high-resolution photographs (and a corresponding full-length documentary film) provided the viewer with documentary "proof" of the objects *in situ*. The underwater depths were reasserted in each gallery of the exhibition, present in the blue glow emitted by backlit lightbox photographs that

A SURREALIST FAMILY HAS THE NEIGHBORS IN TO TEA

Figure 10.7 Carl Rose, "A Surrealist Family Has the Neighbors to Tea," in *The New Yorker* (January 2, 1937). Rose Carl, The New Yorker © Conde Nast.

displayed divers as they retrieved the *Unbelievable's* cargo upon the seabed (Figure 10.8, see plate section). Indeed, one of two deluxe exhibition catalogues was intended to evoke Jacques Cousteau's popular dive books.[29]

The show contained a room of vaguely Surrealist mannequin-esque female nudes – Disney-fied, bedazzled with barnacles and lurid coral outcroppings. They were shown beside a Photoshopped installation photograph of the 1936 Surrealist exhibition in London – the very scene of Dalí's gallery-dive. In the altered 2017 image, a large plinth displaying Hirst's five nude torsos was inserted to scale into the London gallery, directly in front of a wall containing paintings by Max Ernst. However, this reference to historical Surrealism went largely unacknowledged, as if the presence of this uncanny imagery was so obvious as to need no explanation. By way of (fictional) justification, a single sentence in the exhibition pamphlet claimed that reproductions of the sculptures circulated among the Surrealists. However, that "Surrealism" was among the historical objects on display both underscores the current of maritime salvage at hand in the present study, as well as confirms the irrevocable sea-change wrought upon Surrealism by processes of commodification.[30] While I hesitate to analyze Hirst's unrelentingly surface-level exhibition-spectacle too deeply, his Surrealist homage – whether it was conscious or not – is a fitting means of conclusion (indeed, to call Dalí-as-diver something like Hirst's spiritual forefather would be to give Hirst too much credit.) Hirst's salvage in and of Surrealism in an entirely fabricated shipwreck and elaborately staged maritime recovery operation illuminates, above all, the perilous, distorted conditions of the present market for art in the age of global capital.

In the diver and the shipwreck, Breton and his fellow adventurers provided insight into the individual and collective anxiety of an age. Between 1919 and 1939, Breton worked to assimilate the Surrealist work of exploration and interpretation of the unconscious qua Freud to an increasingly Hegelian, dialectical materialist inquiry grounded in Marx. This paper has examined this shift through the evolving trope of the diver and shipwreck in André Breton's literary output, recurring from his early texts on automatism and dream to his materialist analyses of Surrealism's aesthetics and politics, and the historical collective unconscious. Yet we can recall that Breton took the shipwreck as fact. Asked about the future aims of Surrealism in an interview in 1954, Breton replied, "Que peut-il demander de plus que d'être cette bouée phosphorescente dans le naufrage?"[31] Nothing can or has been spared. Breton and Dalí have descended into the depths of history, but flash up, phosphorescent and resplendent with verdigris and rust.

Notes

1 André Breton, *Soluble Fish* (1924), in *Manifestos of Surrealism*, trans. Richard Seaver and Helen R. Lane (Ann Arbor: University of Michigan Press, 1972), 63–4.

2 Several studies have explored maritime themes within surrealist literature and art. See especially Michael Beaujour, "De l'océan au château: mythologie surréaliste," *The French Review* 42 (1969), 353–79; Marguerite Bonnet, *André Breton, naissance de l'aventure surréaliste* (Paris: José Corti, 1975); Abigail Susik, "Surrealism and Jules Verne: Depth of Subtext in a Collage by Max Ernst," in *Surrealism, Science Fiction and Comics*, ed. Gavin Parkinson (Liverpool: Liverpool University Press, 2015), 16–39; Lauren Fretz, "Surréalisme sous-l'eau: Science and Surrealism in the Early Films and Writings of Jean Painlevé," *Film and History* 40, no. 2 (2010): 45–65; Mark Dion, Sarina Basta, and Cristiano Raimondi, *Oceanomania: Souvenirs of Mysterious Seas: From the Expedition to the Aquarium: a Mark Dion Project* (London: MACK, 2011); Donna

Roberts, "Surrealism and Natural History: Nature and the Marvelous in Breton and Caillois" in *A Companion to Dada and Surrealism, ed. David Hopkins* (Chichester: John Wiley & Sons, 2016), 287-303; Marion Endt-Jones, "Between *Wunderkammer* and Shop Window: Surrealist Naturalia Cabinets" in *Sculpture and the Vitrine*, ed. John Welchman (Aldershot and Burlington: Ashgate, 2013); Ann Elias, "Sea of Dreams: André Breton and the Great Barrier Reef," *Papers of Surrealism* 10 (2013): 1–15; Margaret Cohen, "Underwater Optics as Symbolic Form," *French Politics, Culture & Society* 32 (2014): 1–23. Insofar as the movement possessed a historical memory, it is helpful to begin in setting out the panorama of Surrealists *avant le lettre*. Surrealism's call to adventure possessed several elements characteristic of the "romantic attitude" as it was defined by W.H. Auden, with the assumption that the sea is the site of the real, and voyage the "true condition of man." Insofar as this navigation was inherently fated, the diver-shipwreck configuration clearly operated through intertextual references to Surrealism's romantic and symbolist precedents, which included Samuel Taylor Coleridge's *Rime of the Ancient Mariner* (1798), Edgar Allan Poe's ill-fated voyage to the limit-space of the globe in *Arthur Gordon Pym* (1838), Charles Baudelaire's deathly Parisian sea shanties, Jules Verne's fantastic voyage in *Twenty Thousand Leagues Under the Sea: A Tour of the Underwater World* (1870), Rimbaud's drifting, sinking vessel in *The Drunken Boat* (1871), and Stéphane Mallarmé's foretelling of wreckage in *A Throw of Dice* (1897). In this genealogical mode, Henri Behar, Michel Beaujour, and Marguerite Bonnet have analyzed the polysemous, transformational function of maritime imagery in Breton's writing, highlighting the rich inheritance of Symbolist poetry. More recently, Abigail Susik has examined a Vernian "orientation" within the cultural production of several Surrealists, arguing that this legacy extended beyond explicit citation and entered the realm of subtext, forming a foundational (if at times latent) thematic, structural aspect of the Surrealist *au-delà*. However, Breton's conception of the sea also exceeded the realm of literary metaphor, evincing a potent, fluid interplay not only between text and image, but also between contemporary visual and popular culture and historical events. Studies have highlighted perhaps the most direct example of the intersection of Surrealist visuality with discourses of modern science and technology in the films of Jean Painlevé. Mark Dion, Donna Roberts, and Marion Endt-Jones have addressed the ecological imperatives of natural history, examining the place of oceanic specimens within a culture of the Surrealist "marvelous." It has only been in recent years that the surrealist space of the sea has been approached from the "blue humanities," as Ann Elias and Margaret Cohen have considered the function of underwater photography and film in Surrealism text and film, as well as the inherent surreality of representations of the maritime environment. And yet couldn't one argue that the geographical and conceptual reconfiguration of the Mercator map in "The World at the Time of the Surrealists," published in the Belgian journal *Variètes* in 1929, prefigured the historiographical project at the very heart of the blue humanities and thalassography, as we set out to reorient landlocked histories?

3 André Breton, *Break of Day*, trans. Mark Pollizzotti and Mary Ann Caws (Lincoln: University of Nebraska Press, 1999).

4 "And if you should die, are you not certain of reawaking among the dead?" Breton, *Soluble Fish*, 40.

5 Allan Sekula, *Fish Story* (Rotterdam: Witte de With Center for Contemporary Art and Richter Verlag), 1995.

6 W.H. Auden, *The Enchafèd Flood: or, The Romantic Iconography of the Sea* (New York: Faber & Faber), 1985.

7 For an explication of the interwar period, see Adam Tooze, *The Deluge: The Great War, America and the Remaking of the Global Order, 1916–31* (New York: Viking, 2014).

8 Paul Valéry, *The Outlook for Intelligence*, trans. Denise Folliot and Jackson Mathews (New York: Harper and Row, 1963). While Valéry's shipwreck functioned as a totalizing metaphor with which to issue a warning to an unquestioned modern subject, it relied upon a certain theoretical distance. The Surrealist generation, on the other hand, harnessed the trope for its capacity to disintegrate the very notion of the subject even further.

9 Max Ernst, "Où va la peinture?" response to inquest in *COMMUNE* 21 (1935), quoted in Henri Béhar, *Le Surréalisme dans la presse de gauche, 1924–1939* (Paris: Éditions Mélusine, 2012), 956. Translation my own.

10 Letter dated July 5, 1925. André Breton, *Lettres à Simone Kahn* (Paris: Gallimard, 2016), 256. Breton's allusion to the Williamson brothers' expedition is noted in André Breton, *Oeuvres complètes*, eds. Marguerite Bonnet and Etienne Alain-Hubert, vol. 4 (Paris: Gallimard, Bibliothèque de la Pléiade, 2008), 349/1264 (fn).

11 André Breton, *Surrealism and Painting*, trans. Mark Pollizzotti (Boston: MFA Publishing, 2002), 1. "Surrealism and Painting" was published in installments, beginning with the fourth issue of *La Révolution Surréaliste* in 1925, before it was compiled for publication in 1928. The definitive edition, edited and expanded by Breton, was published by Gallimard in 1965.

12 Stanly H. Twist, foreword to Jules Verne, *Twenty Thousand Leagues Under the Sea* (New York: Grosset and Dunlap, 1917), vi. This edition was illustrated "with scenes from the picture play produced and copyrighted by the Universal Film Company."

13 Breton, *Surrealism and Painting*, 2–3. Breton's reference to Williamson's underwater film in the service of Surrealist visuality in 1925 thus predated his reproduction of a Williamson film still in *Mad Love* (1937). Several scholars have analyzed the function of the image, which Breton captioned "The treasure bridge of Australia's Great Barrier Reef," in illustrating the tenet of "convulsive beauty." Taking this into account, the image becomes something like an abiding, latent key to Breton's articulation of Surrealist visuality beginning twelve years earlier.

14 Desnos attributed his propensity for shipwrecks to his solitary childhood "nourished by maritime catastrophes" in the work of Victor Hugo and others. See Robert Desnos, "Confession d'un enfant du siècle," *La Révolution Surréaliste* 6 (March 1926): 18.

15 Marguerite Bonnet noted the disparity between the published "étrangère à notre pensée consciente," on the one hand, and the continuity of the undersea image in the original manuscript of the *Manifesto of Surrealism*: "...viendra toute seule, tant il est vrai qu'à chaque seconde il est une phrase *qui passé impercue au fond de l'eau*." André Breton, *Oeuvres complètes*, vol. 1, ed. Marguerite Bonnet and Etienne-Alain Hubert (Paris: Gallimard, Bibliothèque de la Pléiade, 1988), 1372–3.

16 For an explication of the connection between dream and the metaphorical function of the sea in Bretonian surrealism, see Henri Behar, *André Breton, ou surréalisme même* (Lausanne: L'Age d'Homme, 1988).

17 Andre Breton, *Second Manifesto of Surrealism*, 159.

18 Jacques Prévert, "Mort d'un monsieur," *Un Cadavre* (1930): 1.

19 André Breton, André Parinaud et al., *Conversations: The Autobiography of Surrealism*, trans. Mark Polizzotti (New York: Marlowe & Company, 1993), 120–1. Breton continued, "The tempest was all around, but it was also within. It was completely outside the circuits in which adventure had hitherto been sought."

20 Pierre Mabille, "Preface in Praise of Popular Prejudices," trans. Michael Richardson and Krzysztof Fijalkowski, in *The Surrealism Reader: An Anthology of Ideas*, eds. Dawn Ades and Richardson (London: Tate Publishing, 2015), 32–40.

21 André Breton, "Speech to the International Congress of Writers for the Defense of Culture," in *Manifestos of Surrealism*, trans. Richard Seaver and Helen R. Lane (Ann Arbor: University of Michigan Press, 1972), 239. Breton's "Speech to the

International Congress of Writers for the Defense of Culture" (1935), in *Manifestos of Surrealism*, 234-42; and "Political Position of Today's Art" (1935), in *Manifestos of Surrealism*, 212-34 were both integrated into Breton's *The Political Position of Surrealism* published in October 1935 (Paris: Éditions du Sagittaire, 1935). "Political Position of Today's Art" was first delivered in Prague in April; his "Congress" speech was delivered in June of that year.

22 Bill Brandt, "Au cimitière des anciennes galères," *Minotaure* 6 (Winter 1935): 4. Brandt took the photographs when he travelled to Scilly in August 1934.

23 An 1898 *Country Life Illustrated* feature described Scilly as a "tangled maze of peril" and noted: "Silent witnesses of former disasters are to be found in Valhalla at Tesco, where the figureheads of many unhappy ships have been collected." *Country Life Illustrated* 3 (March 19, 1898): 330. Augustus Smith, lord-proprietor of Scilly, established the collection of figureheads in the 1830s, and assembled them on the grounds of his estate, formerly the site of a medieval abbey. The Valhalla Figurehead Collection can still be visited, and is now under the care of the National Maritime Museum in Greenwich.

24 Oh the top left, Brandt captures the strange meeting of two incongruent figureheads representing a dolphin and a shepherdess. The three-quarter length male figurehead on the bottom left was salvaged from the *St. Vincent*, a British merchant ship that wrecked while returning from the West Indies with over 600 tons of sugar in 1882. In the *Minotaure* spread, Brandt identified the figure as "Professor Mincarlo," referring to the distinctively-shaped "Mincarlo" rock on which the ship ran aground.

25 Pierre Mabille, "Preface in Praise of Popular Prejudices," 32–40.

26 The wine glass with spoon recalls the glasses of crème de menthe that adorned Dalí's *Aphrodisiac Jacket*, and may be a nod to Buster Keaton, who placed his trademark hat atop a diving suit in *The Navigator* (1924.) Dalí himself attributed much of his international acclaim to the "violent eccentricity" of his diving suit lecture. The fiasco received considerable attention in the American press, and was reported in *Time, The New Yorker*, and *Newsweek*. The details of the story change throughout these accounts: in some, David Gascoyne freed Dalí from his helmet while others insist it was Edward James; *Newsweek* reports that Dalí's wife, Gala, had the key to the brass collar of the suit but momentarily lost it. A 1937 article in *Vogue*, clearly referencing the episode, informed readers that "a surrealist is a man who likes to wear a diving suit, but does not dive." Dalí would capitalize on the image of the Surrealist diver even further with "The Dream of Venus" Pavilion, erected at the 1939 World Fair in New York. The façade of Dalí's funhouse resembled a coral formation.

27 The exhibition was funded and developed in collaboration with the Pinault Foundation, and occupied both Pinault museums in Venice. *Treasures from the Wreck of the Unbelievable*, curated by Damien Hirst and Elena Geuna at the Palazzo Grassi and Punta della Dogana, April 9–December 3, 2017.

28 According to the fabricated history set out in the exhibition, the collection once belonged to a fictional freedman-turned-collector by the name of Cif Amotan II (who, as we glean from portrait busts and an especially heinous barnacled sculpture depicting the ancient collector holding the hand of his friend Mickey Mouse, is in fact Hirst himself).

29 Damien Hirst, Christoph Gerigk, and Steve Russell, *The Undersea Salvage Operation: Treasures from the Wreck of the Unbelievable* (London: Other Criteria, 2017).

30 Damien Hirst, with foreword by François Pinault, introduction by Martin Bethenod, and contributors Simon Schama, Henri Loyrette, and Frank Goddio, *Treasures from the Wreck of the Unbelievable* (London: Other Criteria, 2017).

31 Francis Ponge, "Entretien avec Breton et Reverdy," in *Le grand recueil, II. Méthodes* (Paris: Gallimard, 1961), 287.

11 The shipwreck as undersea gothic

Margaret Cohen

The Western imagination has been fascinated with the catastrophe of shipwreck from the time we land-adapted mammals first to set out to sea. For millennia, this fascination took the form of horrific images of inaccessible depths, signifying punishment for humans venturing beyond their cosmically drawn limits and transgressing the divine order.[1] In the early modern era, the theological suspicion of seafaring yielded to a view of the seas a frontier for empowered agency: for exploration, profit, and knowledge. However, while this agency unfolded over the surface of the seas, "[t]he deep sea's floor," in Steve Mentz's words, remained "as unreachable to early modern Europeans as the moon ... When the sea-bed gets invoked ... it represents the impossible fantasy of knowing the unknowable, reaching the bottom of a bottomless place."[2] Down into this bottomless place, the early modern imagination cast a persistent sense of the audacity of venturing over a planetary realm inimical to human physiology. The depths were figured as a site of horror, where corpses and destroyed technology fed alien creatures. Thus, Clarence's dream in Shakespeare's *Richard III* (c. 1592): "Methoughts I saw a thousand fearful wrecks;/Ten thousand men that fishes gnaw'd upon," scattered on "the slimy bottom of the deep." Ornamenting this horror were materials that would have been vast riches on land but that were useless in this inhuman realm, highlighting the wreck of ambition: "[w]edges of gold, great anchors, heaps of pearl,/Inestimable stones, unvalued jewels."[3]

Maritime technology would finally access the last frontier of the ocean depths in the middle of the nineteenth century. New access was opened by oceanography and observation-based marine biology, along with industrial technologies for diving, devised for marine salvage and engineering.[4] The closed helmet diving suit, for example, developed into a usable pervasive practice thanks to engineering innovations in the 1830s by British engineer/diver John Deane and Prussian-British engineer Augustus Siebe. A milestone in its use was the salvage in the later 1830s of cannons from the HMS *Royal George*, a naval ship that sank in 1782 off Portsmouth, killing over eight hundred seamen.

As the salvage of these cannons illustrates, the new capacity to access the undersea not only opened the most forbidding frontier on our planet, but enabled people finally to observe the wreckage of ships on the ocean floor. Both the new access to the depths, and the observation of physical wreckage began the transmutation of the age-old topos of shipwreck as a horrific grave. This article examines a new vision created by innovators in underwater film during the 1940s and 1950s, a revolutionary period in the ability to film the ocean depths thanks to the invention both of scuba and of new camera technologies. This vision was wreckage shown as a gothic spectacle, emerging as the technological ability to view and record wreckage on the seabed adapted a Western aesthetics of ruin dating to the Enlightenment.

The gothic spectacle of the shipwreck does not appear in Jules Verne's *Twenty Thousand Leagues Under the Sea* (1870), the first novel to imagine the undersea as a holistic planetary environment based on the new access developed across the middle decades of the nineteenth century. At the time the novel was written, the only direct observation of wreckage was done by salvage professionals, who publicized little of what they saw below. Verne hence submerged into the depths material brought up by salvage diving. Catching views of sunken ships, the narrator Professor Aronnax saw beneath the surface "hulls of shipwrecked vessels that were rotting in the depths.... cannons, bullets, anchors, chains, and a thousand other iron materials eaten up by rust."[5] Verne also submerged the popular Enlightenment topos that Hans Blumenberg calls "shipwreck with spectator": the ambivalent experiences of observers watching the doomed from a place of safety, sympathizing with the victims' anguish, chilled by their inability to intervene, and at the same time, relieved that they are safe.[6] Thus, Professor Aronnax paints with pathos the attitudes of the drowned in another ship very recently sunk ("some few hours"). The tableau includes the body of a mother who

> [i]n one despairing effort ... had raised her infant above her head, poor little thing! whose arms encircled its mother's neck.... four sailors ... frightful, distorted ... by their convulsive movements, whilst making a last effort to free themselves from the cords that bound them to the vessel

and a "steersman alone, calm, with a grave clear face, his grey hair glued to his forehead ..."[7]

Compare Verne's visions of wreckage that he devised submerging terrestrial knowledge and conventions to the painting *Davy Jones's Locker*, by the celebrated British maritime painter William Lionel Wyllie, first exhibited at the Royal Academy in 1890, twenty years after Verne's novel was published (Figure 11.1, see plate section). By this time, amateur observers were starting to take vision beneath the sea surface. According to the Greenwich National Maritime Museum website,

> Wyllie made studies for the picture while cruising with his family in the Firth of Clyde in the summer of 1889, going under water with a diving helmet improvised from a biscuit tin. His wife Marion Wyllie later wrote

> of the 'beautiful, green, transparent water, the loom of a wreck in the background, while a big rusty anchor, bones and jewels sprinkle the bottom. Many fish are swimming about, and an octopus, a careful study of which was made in the Brighton Aquarium, lurks in a corner.'[8]

In Wyllie's painting, the décor beneath the waves recalls the early modern imagination, with an anchor and skulls on the sea floor, and in the background of the painting, an evocation of the catastrophe that led to such loss. Yet the subtle rich palette of undersea color, the atmospheric haze, and the bizarre, remarkable contours of marine flora and fauna undulating in the surge imbue the scene with an alluring atmosphere of melancholy fascination. *Davy Jones's Locker* invites us to linger at a site that is at once a grave and yet that also offers captivating visual experience, as well as an image of natural regeneration. The image defangs what might be the unbearable representation of human remains by reducing them to skulls and bones, age-old signifiers of mortality. The dead also receive an afterlife in silver schools of fish and translucent jellyfish amidst the haze, both rendered in the pale colors commonly associated with ghosts. Human remains moreover become a new refuge for marine animals, such as the octopus, that entwines itself among the bones, almost the cerebral anatomy of a cross-species assemblage. Indeed, this term, assemblage, was used by the *Times* critic reviewing the painting when it was first shown, as the critic attempted to put into words the bizarre yet compelling quality of the scene:

> Mr Wyllie has been for a Challenger expedition on his own account [in descending in to the ocean depths to gather data], and none can say that he has not found his reward in the strange assemblage of forms and colours represented in his most original picture.[9]

Davy Jones's Locker in short turns wreckage from a site of horror to a beguiling ruin that is an occasion for fascinated and somewhat eerie contemplation. In Enlightenment and romantic aesthetics, the picturesque was the category used to stage the beauty of ruins, where irregular, disordered and sometimes wild landscape enhanced the contrasts of once majestic, now dilapidated structures. Thus J.M.W. Turner rendered Tintern Abbey in the 1794 view in Figure 11.2 (see plate section). Its soaring gothic vaults leave piles of rubble, while pleasingly irregular foliage ornaments arches now open to the sky. Looking to the back of the scene, the remaining stones of gothic windows seem almost about to turn from stone to air and float off into the atmosphere. And yet, Wyllie's underwater picturesque has a disquieting quality contrasting with Turner's watercolor. This disquieting quality derives from vistas of a world where humans are out of place. The wreck of their technologies expresses their unsuitability to marine conditions, and yet the place is mesmerizing, with its "strange" (to quote *The Times* critic) unprecedented optics, colors, and forms.

The Times critic's analogy between the four-year Challenger expedition pioneering modern oceanography and Wyllie's biscuit tin and visit to the

Brighton aquarium is overblown – but Wyllie was indeed opening a new fantasy of shipwreck based on direct observation of underwater conditions. His fantasy would be amplified with the revolution both in diving technologies and underwater recording techniques in the 1940s and 1950s. In the realm of diving, new technologies included scuba, perfected in the early 1940s. Scuba, an acronym for **s**elf-**c**ontained **u**nderwater **b**reathing **a**pparatus, enabled the diver to swim through a wreck's compressed crumbling quarters with less danger of fouling his or her air supply than that run by helmet divers dependent on hoses from the surface. In the realm of recording technologies, the first legible underwater photographs date to the 1890s and the first underwater film footage was shot in 1914. During the Second World War and immediate post-War era, engineers and underwater explorers worked together on strong artificial lighting systems and lenses adapted to the low-light conditions beneath the sea. The invention of faster color film around 1950 enabled photography and film to portray the beauty of underwater color.

Wreck diving and underwater film came together when French Navy Lieutenant Jacques-Yves Cousteau chose wreck sites as the subject of his first film shot with scuba in 1943, *Sunken Ships* [*Epaves*]. In this black and white twenty-eight minute feature, Cousteau together with fellow dive pioneers Frédéric Dumas and Philippe Tailliez showed their exploration of wrecks on the bottom of the Mediterranean, between Marseille and Saint Raphaël.[10] Testing their new inventions, the divers filmed wrecks as deep as sixty-two meters, beyond the reach of recreational divers even today. Sites of wreckage were only rendered more atmospheric when Cousteau included a color sequence on wreck diving in *The Silent World* [*Le Monde du silence*] (1956). This film earned Cousteau the Palme d'Or at the Cannes Film Festival that year, together with the film's co-director Louis Malle, who started his career working on *The Silent World*, dropping out of the prestigious IDHEC (Institut des Hautes Etudes Cinématographiques) for a job as Cousteau's underwater cameraman.

When Cousteau and his fellow filmmakers and divers exhibited wrecks, they had to invent the aesthetic contours of an environment that had never been seen before. To cite Malle on his experience filming *The Silent World*, "[w]e had to invent the rules – there were not references; it was too new."[11] This challenge was a remarkable opportunity for environmental aesthetics of the marine; one whose closest European analog was representations of hitherto unknown seas in the age of global exploration. Like the artists of this earlier era, underwater filmmakers brought together pre-existing aesthetic conventions with the unprecedented qualities of a new environment. In both *Sunken Ships*, and then in the shipwreck sequence of *The Silent World*, the filmmakers followed Wyllie in conceiving of wreckage as a space of beguiling ruin. Rather than a picturesque site, however, Cousteau and his fellow filmmakers organized the wreckage as a gothic spectacle.

Describing his experiences wreck diving in the book *The Silent World* (1953), which was an inspiration for the film, Cousteau called the wreck of the four-thousand ton freighter *Le Tozeur* "a fine movie studio."[12] According to Cousteau's description, this set recalled the décor of a gothic novel, where ruined abbeys and castles were often the settings for highpoints of drama and revelation. Cousteau called the ship an "iron abbey," with narrow passages and an "intricate stairwell." *Le Tozeur* had "bulkhead openings ... arched in the fashion of cloisters" and "light filtering down as though from clerestory windows." This hazy atmosphere of mystery was ornamented by "[w]eeds that grew like lichens in a damp chapel," as marine life took over once useful technologies. As is common in gothic settings, chiaroscuro created an atmosphere of mystery and menace. Divers could look from "a shadowy gangway at remote blue lights opening on the sea [but] [w]e did not feel ready to swim through the tunnels to those lights."

Cousteau's citation of gothic conventions in both *Sunken Ships* and *The Silent World* extends from the atmosphere of the wreck site to the narrative framing its exploration. The gothic novel, invented in the eighteenth century, takes readers into a haunted world after assuring them that they live in an era of the Enlightenment, when the barbarism of feudal society and the intolerance of medieval religion no longer obtain. In *Sunken Ships*, the topside prelude explains dive and film technology. In *The Silent World*, the topside prelude shows the systematic processes used by the *Calypso* to locate the wreckage, scouring the area where the ship is known to have sunk and marking depths with the aid of sonar. The mood is one of Enlightenment optimism, as the crisp, methodical search unfolds under the bright sun shining on the deck of the marine research vessel. Reminders of modern technology pervade even the soundtrack of the film in the charting sequence: the backdrop of a ticking clock and the hum of the ship's motor.

Now this mood of Enlightenment optimism could conceivably extend to the portrayal of the wreck, as it was an immense achievement to locate a shipwreck and bring it to light. Further, the location of the wreck awakens it. In *Sunken Ships*, the ship will be lifted from the depths and melted down. In *The Silent World*, Cousteau and his team show how scuba and film propel the development of nautical archaeology. Chiseling off the overgrown bell of the wreck, Cousteau and his divers identify the ship as the *Thistlegorm*, a merchant ship built in Britain in 1940. The ship transported refurbished wartime materials to British troops in Egypt and Libya, until it was sunk in 1941 by a German airstrike in the Red Sea off the coast of Egypt. Cousteau brought some of the *Thistlegorm*'s artifacts to the surface and it is now one of the most dived wrecks in the world. In the words of a dive website, it is "a veritable underwater 'World War II Museum.'"[13] The wreck's accessibility would further expand in 2017, with the creation by its archaeologists of a virtual website allowing a world-wide public to tour it remotely.[14]

And yet, when Cousteau's divers descend to explore the *Thistlegorm*'s wreckage, *The Silent World* switches from the Enlightenment mood characterizing its topside process of location. Cousteau and Malle imbue the site with a tone of fascination, menace and ghostly haunting, befitting the vessel's tragic end. More specifically, they organize the wreck diving episode of *The Silent World* in a fashion recalling the conventions used by gothic narrative to take viewers from a present ruled by reason to a tortured, haunted past. As epitomized in novels by the queen of gothic suspense, Ann Radcliffe, the menace of that past crystallizes in scenes where a protagonist feels her reason assaulted as she explores the nooks and crannies of a ruined medieval abbey or fortress, encounters relics – and, despite herself, fantasizes about the return of the dead. One scene in Radcliffe's *The Mysteries of Udolpho* (1794) showing the gothic quality to Cousteau's description of *Le Tozeur* occurs after the genteel, reasonable heroine Emily has been forcibly carried off to a remote castle in the Apennines. She is drawn by a dubious servant into a night-time exploration of a medieval fortress with "almost roofless walls," and "the gothic points of the windows, where the ivy and the briony had long supplied the place of glass." Wind whistles through "narrow cavities," includes winding staircases, vaults and narrow passages "dropping with unwholesome dews."[15] Such architecture confuses Emily and adds to her apprehension about the violence and ghosts lurking in this "obscure and terrible place."

The diver on the *Thistlegorm* in the film *The Silent World* similarly explores nooks and crannies, swimming along the abandoned deck of a ship once full of life. And yet, spectators witness in this sequence a new iteration of the gothic – where the menace comes not only from the ghostly residues of the past but also from the physical qualities of the environment. The classic gothic, both in literature and the arts, sought to explain violent, irrational social power, in an era proclaiming the rule of reason. In both *Sunken Ships* and *The Silent World*, the threat to human life comes from the underwater atmosphere. Most episodes of *The Silent World* exalt technologies of undersea access. The scenes of wreck diving are a pause in such exaltation to remind us of the dangers of this newly accessed planetary realm.

The wreck setting in the film exemplifies how Cousteau and Malle use the physical quality of the environment to reinvent gothic setting. In both *Sunken Ships* and *The Silent World*, the hull of the wreck lies off-kilter on the ocean floor, and the camera portrays its tilt. The portrayal is particularly dramatic in the lush color of *The Silent World* (Figure 11.3, see plate section). In topside cinema, such an off-kilter shot is called a Dutch angle or canted shot, "composed with a camera tilted laterally, so that the horizon is not level and vertical lines run diagonally across the frame."[16] Gustave Mercado credits the invention of this shot to German expressionism, pioneering the use of "spatial imbalance or disorientation" to suggest "dramatic tension, psychological instability, confusion, madness, or drug-induced psychosis." As developed in subsequent terrestrial cinema, the shot is used to portray a "character's altered or abnormal state of mind ... a stressful or unusual situation," or "to convey that an unnatural or abnormal situation is taking place." The canting of the *Thistlegorm* is in fact how the wreck settled on the ocean floor. The trained Navy divers of the *Calypso* do not share the distorted perspective associated with the canted shot above land: they explore the *Thistlegorm* calmly, using rational technology essential to survival undersea. And yet, the emotional resonance of the canted shot at the same time fits the scene. It evokes the altered phenomenology of perception in the depths, the horror of the lives lost beneath the waves, and the bizarre (by terrestrial standards) physical features of the setting.

The undersea gothic also transforms the figure of the gothic explorer, as s/he makes his or her way through landscapes haunted by the past. In classic gothic, spectators identify with a protagonist like Radcliffe's Emily, sharing her fear and determination. While viewers identify with the determined crew of the *Calypso* as they search for the wreck and share their excitement in its discovery, identification with the diver becomes more tenuous. The film does not give us his thoughts through a voiceover, and his countenance is hidden by his mask. Further, when he glides through spaces where people once walked and lived, his smooth, floating underwater movement recalls long-standing fantasies of ghosts' locomotion. The diver's sinuous gliding is both ghostly and environmentally motivated, like the canted angle of the wreck. Such movement is a hallmark of efficient swimming and also helps the diver avoid contact with the crumbling surfaces of the vessel. In the book *The Silent World*, Cousteau specifies, "[t]he *Tozeur* was as treacherous as she was receptive, and taught us much about the actual perils of wrecks," from the organisms that could injure the diver to the danger that the wreck would collapse.[17] In the film, viewers do and do not identify with this post-human figure silently gliding by, a disturbing tension accentuated by the contrast between image and soundtrack. While the diver moves as if he were a ghost or fish, the anxious score gives way to the human sounds of the diver breathing through his regulator, as he starts to explore the wreck's interior.[18]

The exhibition of terrestrial technologies stripped of use emphasizes the alien realm spectators visit in *The Silent World*'s undersea gothic. Thus, the windows and portholes become meaningless in an environment where the distinction between the protective inside and the outside has been destroyed. In Figure 11.4 (see plate section), we see fish looking through what was once a window onto other fish beyond. (c. 43:18). We catch haunting glimpses of once useful elements that have lost their function in tragedy, like the coil of rope dangling from the edge of the ship in Figure 11.3. It resembles a hangman's noose, evoking both the drowned and the danger to the diver from the wreck environment.

While Verne imagines the faces of the drowned, such horror is eschewed in Cousteau's exhibition of the wreck as gothic spectacle.[19] This stance may be in keeping with Cousteau's exaltation of the new human access to the undersea, which dedramatizes its danger. Cousteau's stance as demystifier may also explain the appeal for him of gothic spectacle as a way to organize wreckage. The gothic traffics in ghosts as aesthetic phenomena, using poetics and sensibility to hold them at bay so that their stories can enrich but not overwhelm the present. The aesthetic approach to the dead is evident in the visit to the *Thistlegorm*, where instead of remains, we see melancholy traces of humanity, like the useless window or rope, or supplies like lanterns toppled inside the ship.

Throughout the classic gothic, there are charged moments at the climax of the drama when the dead spring to life – whether in the imagination of the observer or as actual ghosts, when the story takes a supernatural turn. Such an electrifying moment of contact between the present and the past is found in Cousteau's undersea gothic as well. Already in *Sunken Ships*, and then in *The Silent World*, the present and the past converge through touch, in contrast to the diver's general aversion to brushing against the wreckage. In *Sunken Ships*, there are two moments of contact: the diver grasping the iconic eight-spoked wheel that, "[a]n old helmsman took ahold of . . . for its first launch," as well as finding a tube that "was a canon. The wheel to aim it still turns."[20] In contrast to such voiceover, *The Silent World* brings present and past into contact without explanation. When the diver finds the ship's bell, before reading the inscription, he strikes this meaningful artifact twice, awakening it from its long sleep. Nautical archeologists use the bell for identifying a vessel, as is the case in *The Silent World*, when a disembodied female voice reads the legend: "Thistlegorm Glasgow" (c.37:39–42).[21] Ships' bells are "the 'symbolic embodiment of the ship itself,' . . . used to mark the passage of time onboard the vessel, struck every half-hour, day or night, as well as to signal the change of the crew's watches."[22]

In a recent insightful article on gothic temporality, Jesse Molesworth asks why gothic fiction, set in the religious past, is fascinated with modern clock time, shown in the genre's fondness for the fatal hour when dire events occur. In Molesworth's explanation, the gothic preoccupation with the hour offers a kind of sacred time for a secular age. The homogeneous hours of the clock become qualitative: "not mere witnesses" to plot but rather they "each possess an almost talismanic power."[23] Molesworth's point is relevant to making sense of the moment when the diver strikes the bell of this ship, in a sequence following the work of locating the wreck accompanied by the sound of a ticking clock. I'd suggest the talismanic power of the bell here affirms human existence beneath the sea, but also acknowledges its ambivalent, spectral overtones. This ambivalence is also found in the voice that then identifies the ship, reading text on the bell. This female voice speaking with a British accent is an unidentified presence in a film entirely peopled with male characters, also narrated entirely in French. Nor was a woman among the *Thistlegorm*'s crew or the nine who died there. This voice devoid of referent speaks through air, which is also the sound quality of the bell tolling. The disjunction between location and recording enhances our sense of humans as at once present and out of place.

At the same time, Cousteau and Malle's exhibition of wreckage shares with Wyllie's *Davy Jones's Locker* a creative element intimating a new form of life. *The Times* critic used the term assemblage to describe the originality of Wyllie's painting in 1890. The artistic movement pioneering assemblages, works crossing materials and categories, was surrealism, although the term assemblage would not be applied to art in English until 1961, according to the *Oxford English Dictionary*.[24] At the same time, the surrealist art of assemblage resonates with the content of Wyllie's scene, even if Wyllie used the single medium of oil. Surrealist assemblages violate conventional (terrestrial) categories and orders, like Wyllie's painting, albeit through bringing together disparate elements. Surrealism brought such elements together not to call up the past, like the gothic, but rather to usher in a new, hitherto unimagined future. Wyllie's painting too looked towards new experience when it exhibited a mysterious, unprecedented environment beguiling the viewer.[25]

The sense of potential is also evident in Cousteau and Malle's gothic spectacle of the wreck, and in this case too, the content of this future is mysterious and ambiguous. Cousteau struggled to put into words his sense of potential emerging from wreck diving in the book *The Silent World*, echoing the narration of creative contact between present and past in *Sunken Ships*:

> [t]o one who glided easily across the moss-covered deck, nothing was wood, bronze or iron. The ship's fittings lost their meaning. Here was a strange tubular hedge, as though trimmed by a fancy gardener. Didi [Frédéric Dumas] reached into the hedge and turned a wheel. The cylinder rose smoothly. It was a gun barrel.[26]

One famed surrealist example of creative contact between disparate elements includes the meeting of a sewing machine and an umbrella on a dissecting table. The gun barrel that is also a strange hedge on a wreck site bathes in this surrealist atmosphere. However, Cousteau found this scene not in art but in the reality of wreckage beneath the sea.

The spectral potency of undersea wreck sites remains as increasingly sophisticated technology expands the reach of nautical archaeology into new frontiers of the seas. When James Cameron filmed the wreckage of the *Titanic* at a depth of approximately 12,000 feet, far beyond the reach of unassisted diving, he utilized one deepsea submersible for the filming, two others for a custom-designed lighting system, and a remote camera "Snoop Dog," which he maneuvered through the narrow passages of the ship. Despite the extraordinary achievement of this footage, Cameron did not only vaunt the power of his expanded underwater access, but also compared the wreck to a "gothic ruin." He likened exploring the *Titanic* with his ROV "avatar" to a "strange out-of-body experience of ghostwalking."[27] Surrealism was another aesthetic that came to Cameron's mind when he sought to express the creative quality to this "gothic ruin [that] exists now in a ghostly limbo, neither in our world nor completely gone from it. The rusticles have transformed Edwardian elegance into a phantasmagorical cavern, a surreal underworld ruled only by dream logic."

Both the hedge that is a gun barrel and the phantasmagorical rusticles have a lugubrious tone. The image of potential concluding the wreck sequence in *The Silent World*, in contrast, has a tinge of whimsy, another powerful surrealist

Figure 11.6 Jacques-Yves Cousteau and Louis Malle, *The Silent World* (*Le Monde du silence*) (1956). Screenshot of the bubbles rising like jellyfish (c. 43:54).

mood. The gothic exploration conventionally ends with the revelation of a fatal secret – one involving lost ties of birth and family, as well as undead souls that must be put to rest. In the image concluding the sequence on the *Thistlegorm*, in contrast, air bubbles escape through holes in the deck of the sunken ship (Figure 11.5, see plate section). With these bubbles, the last gasps of the drowned converge with the reassuring flow of air from the breathing diver, and they rise towards the lighter atmosphere of the surface water, with their tops swelling and their bottoms flattened, making the bubbles resemble jellyfish (Figure 11.6). With this whimsical assemblage, conjoining wreck, diver, and marine organisms, the *Thistlegorm* sequence ends suggesting the wreck's awakening to what its observers, from Wyllie's critic to Cousteau to Cameron call a "strange" new form of life.

Notes

1 In Hans Blumenberg's words, "there is a frivolous, if not blasphemous, moment inherent in all human seafaring, on a par with an offense against the invulnerability of the earth, the law of *terra inviolata*, which seemed to forbid cutting through isthmuses or building artificial harbors — in other words, radical alterations of the relationship between land and sea." Hans Blumenberg, *Shipwreck With Spectator: Paradigm of a Metaphor for Existence*, trans. Steven Rendall (Cambridge, MA: The MIT Press, 1997 [1979]), 10–11.

2 Steve Mentz, *At the Bottom of Shakespeare's Ocean* (New York: Continuum, 2009), xiii.

3 William Shakespeare, *The Life and Death of Richard the Third*, Act 1, scene iv, cited from http://shakespeare.mit.edu/richardiii/ (accessed May 2018). I thank Steve Mentz for drawing my attention to this passage.

4 Helen Rozwadowski provides a cogent history of nineteenth-century access to the deep sea in *Fathoming the Ocean: The Discovery and Exploration of the Deep Sea.* On the spectacle of the ocean's depths back on land, see, among others, Bernd Brunner, *The Ocean at Home.*

5 Jules Verne, *Twenty Thousand Leagues under the Sea*, ed. and annotated by Walter Miller (New York: Thomas J. Crowell, 1976), 121.

6 Blumenberg describes the Enlightenment topos in *Shipwreck with Spectator* c. 35 ff.

7 Verne, 120.

8 The website continues, "A number of the studies are in the NMM's Wyllie Collection (for fish and octopus studies, see PAD0488-98; for wreck study see PAF0685)." http://collections.rmg.co.uk/collections/objects/561491.html (accessed May 2018).

9 Review of *Davy Jones's Locker, The Times*, May 3, 1890, quoted on http://collections.rmg.co.uk/collections/objects/561491.html (accessed May 2018).

10 Information on *Sunken Ships* on the website dedicated to the memory of Cousteau's co-diver Philippe Tailliez: http://www.philippe.tailliez.net/naissance-du-gers-et-des-premiers (accessed May 2018).

11 Louis Malle, *Malle on Malle*, ed. Philip French (New York: Faber and Faber, 1993), 8.

12 Jacques-Yves Cousteau, *The Silent World*, with Frédéric Dumas (New York: Harper & Row, 1953), 41. Remaining quotes in this paragraph from page 41.

13 "Thistlegorm Dive Site,? http://www.eazydivers.com/eng/index.php/thistlegorm (accessed May 2018).

14 "The Thistlegorm Project," http://thethistlegormproject.com. See "Virtual Tour of Second World War Shipwreck Goes Online," *The Guardian*, October 6, 2017, https://www.theguardian.com/world/2017/oct/06/virtual-tour-of-second-world-war-shipwreck-thistlegorm-goes-online (accessed May 2018).

15 Ann Radcliffe, *The Mysteries of Udolpho* (New York: Penguin, 2008), phrases from 344–346.

16 Gustave Mercado, *The Filmmaker's Eye* (Burlington, MA: The Focal Press, 2011), 101. All quotes about the canted shot from this page.

17 Cousteau, *The Silent World*, 40.

18 The soundtrack of *Sunken Ships*, in contrast, does not offer such disturbing tension between the visual and auditory evocations. Rather, a voiceover recalls romantic couples who promenaded on the decks at night, accompanied to the strains of a waltz.

19 Film critic André Bazin viewed the exhibition of the dead in a purely informational manner without "any moral or aesthetic justification" as obscene, remarking on the drowned countenance of a pilot "as if it were just another curiosity," on a 1956 program on *Journal télévisé*, which had underwater sequences "rather in the style of *Le Monde du silence.*" André Bazin, "Information or Necrography" [1957] in *André Bazin's New Media*, trans. and ed. Dudley Andrews (Oakland: University of California Press, 2014), 125, 124. I thank Dudley Andrews for drawing my attention to this passage.

20 Jacques-Yves Cousteau, dir., *Epaves* [Sunken Ships], Film (France: Jacques-Yves Cousteau producer, 1946 [1943]), c. 8:00 ff. The original French is "un vieux timonier l'a pris en main avec orgueil pour son premier appareillage. Et ce tube c'était un canon. Le volant de pointage tourne encore." My translation.

21 Jacques-Yves Cousteau and Louis Malle, dirs., *Le Monde dusilence*, Film (France: FSJYC Production, 1956). The *Calypso* did indeed locate this wreck, and throughout his career, Cousteau participated in the development of nautical archaeology.

22 Chris Sorenson, "Ottawa Shows Off Bell Recovered from the Erebus,", *Maclean's*, November 6, 2014, http://www.macleans.ca/news/canada/ottawa-shows-off-bell-recovered-from-hms-erebus/ (accessed May 2018). I thank Adriana Craciun for this insight.

23 Jesse Molesworth, "Gothic Time, Sacred Time," *Modern Language Quarterly* 75, no. 1 (2014): 29–55, 38. Molesworth writes, "[t]he gothic clock acknowledges the temporal modernity described by Thompson and others through its sheer ubiquity, but the time discipline it enforces owes less to the rise of industrial capital than to a strongly developed sense of ritual practice" (45).

24 *Oxford English Dictionary*, entry for "assemblage," http://www.oed.com.ezproxy.stanford.edu/view/Entry/11781?redirectedFrom=assemblage#eid (accessed May 2018). At the time of *Davy Jones's Locker*, the term had a range of uses, including in paleontology, archaeology and geology for "a group of artefacts found at the same site."

25 Wyllie's multi-faceted environment is more extensive and inviting than the "sea-change into something rich and strange" of the human remains alone, sung about by Shakespeare's Ariel in *The Tempest*.

26 Cousteau, *The Silent World*, 39.

27 James Cameron, "Ghost Walking in *Titanic*," *National Geographic*, April 2012, vol. 221, issue 4, 100–1. http://www.nationalgeographic.com/magazine/2012/04/titanic-shipwreck-underwater-exploration-james-cameron/ (accessed May 2018). All Cameron quotes in this paragraph from the article.

12 Deep time and myriad ecosystems

Urban imaginaries and unstable planetary aesthetics

Linda Williams

Most of the world's population now lives in cities,[1] and from this urban perspective the planetary scale of the oceans and their myriad, multi-scaled life forms are far removed from everyday experience. Just as the night sky is obscured by the lights of the city, so the common perception of the world's oceans is similarly restricted: not just in response to their vast spatial scale, but also to the almost unfathomable time scales of their ancient origins. In this sense, the spatial and temporal restrictions of the urban view extend beyond the built environment as such, to how the ocean is more generally understood in aesthetic representations, and not least in its reconfiguration in the visual screen culture common to all contemporary urban centres. Conventional representations of the ocean in the city have been to some extent enhanced by recent digital media, especially in revealing a new imagery of the submarine world. Yet film of oceanic life is nonetheless framed by personal or cinema screens in ways that often do not convey a compelling aesthetic sense of its disconcertingly fluid transitions of scale, or mighty, unruly energy.

The oceans as channels of colonisation and international trade poured great wealth into the major cities of the Western world, and were the primary conduit of globalization in the "long seventeenth century"[2] of early modernity. Yet within the everyday social fabric of the city itself, the actual *materiality* of the ocean effectively moved in a countervailing motion: withdrawing, as it were, like a great retreating tide into the abstractions of global capitalism. Moreover, as a primordial ocean abated to human navigation, it was from the late nineteenth century in Peter Sloterdijk's plausible account, when it was sucked ever faster and deeper into what he calls the *world interior* of capital.[3] This has since become a domain of ruthless expedience where marine life became increasingly degraded by gradual global warming,[4] pollution[5] and the insidious effects of ocean acidification,[6] to the extent that it will likely soon cease to feed and nurture cities in the ways it has for so long. Meanwhile, recent research suggests that people in cities are generally unaware of the gravity of the anthropogenic threat to the world's oceans,[7] while at the level of a cultural economy of affects, there are very few aesthetic forms that reveal the crucial role of the oceans in sustaining global ecosystems, or how deeply their material histories are enmeshed in the human story.

In this chapter I aim to show that while the aesthetic experience of oceanic space and time in screen cultures certainly has potential – especially in recent elevated perspectives on the planetary scale of the oceans, there is also a less abstract, more biosemiotic aesthetic of oceanic temporal and spatial scales available to people in modern cities. Though these more concrete urban forms owe as much to the social hierarchies of cultural production as the radiant blue oceans of digital display, they resist the tendency to sanitisation that remains widely evident in the screen aesthetic. Instead, they are imbued with traces of marine residues and a more palpable aesthetic sense of materiality reconnecting the histories of the ocean with terrestrial and urban space.

Screen cultures

One of NASA's most iconic images is the photograph of the earth taken from space in 1972 by the astronauts of Apollo 17, known as *The Blue Marble*. Over the last four decades this image has accrued international social prevalence: not only in realigning the visual sense of the earth as a "blue planet" where the oceans comprise 71 per cent of the world's surface, but also as a compelling icon of the complex ecosystems upon which we all depend. More recently, writers such as the Australian Clive Hamilton, have regarded such distant visualisations of the earth as technocratic abstractions of the lived materiality of earthly life, and as such, as representations more conducive to regulatory political regimes than respect for global ecosystems.[8] In contrast to this view, the French philosopher Michel Serres, on the other hand, sees images of our planet taken from space as visual portents of a new, potentially ethical contract between the earth, science, and the cords of human love and violence that bind those two domains.[9] Yet it is perhaps NASA's more recent photographs and videos of the world at night: "The Black Marble" series, that represent a more incisive response to the fragile ecosystems of the blue planet from the perspective of urban populations (Figure 12.1, see plate section).

Though very few stars can be seen from cities, this series from NASA inverts the urban perspective to provide a nocturnal portrait of the human presence on earth. From this elevation, the oceans are signified by vast expanses of saturated darkness interspersed by continental landmasses revealed in the accretions and chains of light that are the world's towns and cities. It is a perspective from which it is not difficult to imagine the millions of people and urban vitality suggested by these intense fusions of light, arrayed as an ever more tightly integrated global web of communications and commerce.

What is less visible, however, is how such socio-economic webs extend into the black expanses of the oceans that surround them, or how the oceans themselves are reconfigured and perceived within these cities of light. In 2008, NASA developed a new kind of map which bears disturbing resemblances to this satellite spectacle. This is a global map of aquatic dead zones,[10] where the deep water is so low in dissolved oxygen that sea creatures cannot survive there, and they are most densely concentrated in those areas of light in the satellite image: along the Atlantic coast of the US, in the North Sea of Western Europe, and the Yellow Sea of East Asia. Indeed, in the aesthetic field of NASA's nocturnal images of the earth, it is as if the glittering centres of New York, London, or the Asian megacities are somehow frozen in time and spatially dislocated from their direct impact on the biodiversity of coastal seas. It is, however, precisely an aesthetics of temporal rupture and spatial dislocation that I want to suggest has particular relevance to how city dwellers might reconceive the ocean through a slow turn towards a more biotic imaginary and affective aesthetics of scale, arising almost incongruously from the escalating urban pace of late modernity.

In a well-known mid-nineteenth-century reflection on the ocean and time, *Dover Beach*, the English poet Matthew Arnold traced the inexorable withdrawal of the ocean from a world rendered unstable by the forces of modernity:

> Its melancholy, long, withdrawing roar,
>   Retreating, to the breath
> Of the night-wind, down the vast edges drear
>   And naked shingles of the world

In taking a cue from how Arnold's mournfully retrospective, nineteenth-century image of an ocean in recoil may speak directly to the cities of the present, I will now turn to two urban reconfigurations of oceanic creatures representing the largest and the smallest of marine life forms: a spectacular whale and invisible microscopic zooplankton. Both these cultural reconfigurations of the ocean entered the nineteenth-century city as museum exhibits, and remain highly visible examples of how residues of the ocean are filtered and represented in urban aesthetics. The whale refers to human extractions of the upper, photic zone of the ocean in the nineteenth century as a means of illuminating of the modern city. The micro-plankton on the other hand, refer to ancient processes of sedimentation in the submarine depths of the ocean, and how oceanic deep time manifests in the fundamental material substance of the nineteenth-century city and its artefacts. As case studies, their differences in scale and visibility are germane to Stefan Helmreich's point that the ocean has a "fluid capacity to link the smallest microorganisms to the largest ecosystem,"[11] to which I would add that the visual presence of oceanic creatures in the city has the capacity to ignite the public imagination on the fragility of oceanic ecosystems.

The naked shingles of the world

In 1891, a whaling ship in the Irish Sea attacked and relentlessly pursued a huge blue whale. Distressed and badly injured from the pursuit, she was eventually driven into the mouth of Wexford Harbour where her fluid form was forced to beach as an inert mass on the shingled shore. The whale was a mature specimen of her kind, whose massive skeleton was purchased at Wexford by the Natural History Museum in London where it was later exhibited in 1938. Since then, her skeleton has been on display at the museum, and at 28.6 metres is perhaps the largest captive blue whale specimen in the world. At the time the whale was killed, this once abundant species had been hunted almost to extinction. The Blue Whale is the largest animal that has ever lived on earth, and this particular specimen was killed for much the same reason as other whales, which is to say, it was extracted from the waves as a readily available living resource of flesh, oil and bone. Its considerable financial value notwithstanding, however, the Wexford whale came to play a cultural role in how the distant oceans are represented in the city, and one that provides clear outlines of how aesthetic representations have shifted over time.

Whaling clearly has deep historical and cultural connections to coastal communities, yet since the flesh, bone and ivory from whales were put to use in a range of urban products from soap and candles to corsetry and piano keys, fragments of whale bodies were distributed across most early modern European cities. The most valuable product of whaling was the oil required to illuminate early modern cities, effectively rendering whales as unwilling constituents in the development of the urban industrialisation now globally extant in NASA's black marble. From this one blue whale hunted to ground at Wexford, for example, the local people extracted 630 gallons of oil from its carcass before the skeleton was sold to the museum for £250. Whale oil had been used to light the streets of London from the seventeenth century, and as whaling increased, so the value of such oil for streetlights decreased in the eighteenth century.[12] By the early nineteenth century, whale oil for streetlights was gradually superseded by mineral oils, and by the end of the century when the Wexford whale was killed, the oil lamp system was being replaced by the fossil fuel of natural gas – itself later replaced by electricity also derived from fossil fuels in the early twentieth century. So far from simply being the leviathans familiar to rural coastal traditions, from the seventeenth century whales made a significant, if often overlooked contribution to the development of urban modernity.

The massive skeleton of the Wexford whale was put on display in London's Natural History Museum's hall of mammals in 1938, where it has continued to engage urban audiences, especially in response to its sheer size. As Philip Hoare has pointed out, this was also a period when the mass slaughter of whales increased with developments in whaling technology, when in the following year, for example, over 40,000 were killed in the Southern Ocean alone.[13] The whale skeleton from Wexford dwarfed those of other mammals in the collection, a leap in the scale of living creatures including the human which the visitors themselves could hardly fail to notice. The whale loomed large in the hall of mammals as a creature several metres longer than the huge plaster cast of an ancient diplodocus skeleton popularly known as "Dippy" the dinosaur, which from 1979 dominated the main hall of the museum on the ground floor. From 2017, however, Dippy has been replaced by the Wexford whale, renamed as "Hope" and reconstructed in a more dynamic "diving" posture with the aim of re-engaging the public imagination in the abundant, if increasingly fragile, life of the oceans (Figure 12.2, see plate section).

After this change in curatorial strategy, the museum director, Sir Michael Dixon, when interviewed by the BBC, emphasised the need for a dramatic object as the centrepiece of public display, but also remarked on the scientific requirement to exhibit a real object from the natural world, rather than a plaster reproduction.[14] Dixon also outlined three main narratives the whale skeleton offers to the public: the first tells the important story of the evolution of sea mammals, their emergence from the sea and skeletal similarity to land mammals, along with the return of cetaceans and other marine mammals from land back to the ocean. The second is the story of the human exploitation of whales. This has special relevance to this particular specimen, since the events of her death are known, and as such bring an accessible, more nuanced narrative to the wider story in which blue whale numbers crashed from a much-reduced population of around 250,000 at the end of the nineteenth century, to fewer than 2,000 by 1967 when blue whales finally gained the legal status of a protected species. The third story Dixon wanted the museum to convey to the urban public, is the more heartening tale of how blue whale numbers have now grown tenfold from 1967 when they gained their new legal status to a current population of around 20,000. This is still a very small number, and as Dixon commented in an interview:

> This is an important and necessary change. As guardians of one of the world's greatest scientific resources, our purpose is to challenge the way people think about the natural world, and that goal has never been more urgent.[15]

Taken together, the aims of the natural history museum in displaying the Wexford whale represent a significant turn towards engaging the public in a more dynamic biotic imaginary than was once made available from the display of the diplodocus cast. From 2017, people entering the museum in central London immediately encounter a specimen of a wild creature that still lives somewhere "out there" on earth. They are invited to imagine its ocean habitat in order to envisage its life, and thus potentially to make an imaginative leap from city to ocean, if not to the deeper historical connections between them.

Those deeper historical connections have recently been a focus of urban architectural theory in the notion of planetary urbanism,[16] an interesting shift in urbanist discourse towards a model that has long been understood in both environmental philosophy[17] and earth systems science,[18] as the view that human lives in cities are critically interconnected with the complexity of global ecosystems. Nonetheless, since at least the era of classical antiquity there has been an enduring binary spatial logic that continues to shape the idea of the Western city in ways that render concepts of non-urban space or "wilderness," including the oceans, as not only inferior to the spatial and spectral domains of cities, but also as somehow ontologically distinct from them.[19] This spatial logic was always fundamentally flawed, but in an age of global environmental crisis it has become a serious impediment to the understanding of the crucial ecological connections between urban and non-urban space. The Natural History Museum in London has now aimed to make such connections more concrete and publicly compelling with the visual display of this massive creature designed to effect maximum impact on the viewer.

Deep time in the city

There are, however, myriad other much smaller skeletons from the ocean to be found in the city, some of which are so minute they are invisible. And though the slightly larger than life statue of Charles Darwin on the main staircase of the Natural History Museum does not at first appear to suggest that the city has an interdependent material relationship with global ecologies, on closer examination, it has the capacity to open unexpected urban vistas of the oceans across deep time (Figure 12.3, see plate section).

This national cultural artefact is the work of the Victorian portrait sculptor Sir Joseph Boehm, and was unveiled to public acclaim in 1882, two years after Darwin's death, by the Prince of Wales and Darwin's champion Sir Thomas Huxley. The museum itself had opened a year earlier, and was conceived as a new way of bringing science to the people. From the outset, the museum was described as a "Cathedral" or "Temple of Nature,"[20] and was designed by Alfred Waterhouse, an admirer of Romanesque religious architecture and much given to the mid-nineteenth-century taste for medieval imagery that shaped his plan for a robust building encrusted with sculptures of insects, plants and animals.[21] Such references to nature extended to Waterhouse's use of materials, as the façade and sculpted creatures on the walls and columns are made from *terra cotta*, literally: baked earth, while the museum's grand entrance was inspired by the basalt columns of volcanic rock in Fingal's cave in the Scottish Hebrides. In the early twentieth century, Boehm's marble portrait of Darwin was moved to a marginal place in the museum, but by the Darwin bicentenary of 2009, the sculpture was returned to its original, privileged position.[22]

In the nineteenth century, sculptures in marble still carried the cultural imprimatur of classical antiquity, though in this work Darwin appears in secular dress rather than classical or scholarly robes.[23] Seated in a modest chair in a relaxed pose, he is composed as a modern man of his time. Yet the implied allusion to the profound influence of Darwin's thought, coupled with the intense focus of his sculpted gaze, imbues the work with an uncanny potential to seem to gaze across exhibits and visitors alike – even to their common material origins, and hence to the geological concepts of deep time which had greatly influenced Darwin's ideas on evolution.

From his student days under the influence of the geologist Adam Sedgewick, Darwin had become familiar with the idea of geological deep time as expounded by James Hutton and Charles Lyell. As a young man of 22, he had taken four books in his luggage when *The Beagle* departed in 1831 for a five-year voyage[24], one of which was Lyell's recent study *Principles of Geology* (1830). The frontispiece of Lyell's book included an engraving of the Roman *Temple of Serapis* near Naples, where the marble columns had been stained by watermarks and erosion from the marine molluscs that had clustered on their surfaces at a time when the site was submerged below sea level. These columns had *gradually* re-emerged from the sea over the centuries, and as such were graphic correlatives of the geological processes forming strata and fossils which rather than appearing to have been formed in ways coeval with Victorian narratives of biblical creation, were now emerging as the troubling signs of a material *longue durée*.

The nineteenth-century concept of deep time also extends to the materiality of the sculpture itself, since marble is a metamorphosed form of limestone, named from the Greek *mármaros*, a shining stone, a sedimentary rock comprised of layer upon layer of ancient calcium carbonate from the tiny exoskeletons of marine organisms. Sedimentary stone such as limestone and marble has been used as an architectural material for many centuries in buildings as diverse as the pyramids, gothic cathedrals or the Empire State Building in New York City. It is often used today as floor tiles or for the stone fireplaces at the heart of urban domestic dwellings, or the *oikos* at the centre of the human world. Hence if we consider the structure of our cities from one such perspective of deep time, there are invisible yet concrete forces of connectivity between the protective shells of marine organisms and the human architectural "skins" or, as it were, the cultural exoskeletons we construct as one of the fundamental urban relations between our bodies and the world.

The inchoate connection between geological deep history and human cultures was the subject of one of W.H Auden's greatest poems, "In Praise of Limestone," written in 1948 shortly after the end of the Second World War, when "If it form the one landscape," Auden wrote,

> that we, the inconstant ones,
> Are consistently homesick for, this is chiefly
> Because it dissolves in water.[25]

Auden's notion of our homesickness for the malleable qualities of water is cause for further reflection on the oceanic origins of human evolution itself. If limestone is the materialization of slow water, in the context of the early twenty-first century it also offers a point of contemplation on how the ocean as an ancient source of biodiversity is now transformed by the rapid increase in ocean acidification that is eroding the sense of home experienced by countless marine creatures.[26]

Etymologically, the word "home" in proto Germanic meant dwelling or village; in the Old Norse, residence, or world; and in the Greek, the *oikos*, or place of habitation in the world. These are all relational terms for a spatial orientation of the human place in the world; though the anthropocentric Greek model of the *oikos*, or human world as home, was later extended to pertain to other life forms when the German biologist Ernst Haeckel coined the term *œcology* in the mid-nineteenth century to signify relations between organisms, and in turn, between organisms and the wider environmental world.[27] This concept of œcology was taken further still with the German term *umwelten*: the idea of a vast interconnected web of subjective worlds developed by the early twentieth-century biologist and semiotician Jakob von Uexküll.

Von Uexküll referred to a "centrifugal architecture" of organisms[28] with each being at the core of its own perceptual home. For von Uexküll, such worlds constituted the individual locus of meaning for countless millions of diverse creatures, including humans. And as Louise Westling's recent study acknowledges, by the mid-twentiethth century, Merleau-Ponty, following von Uexküll, saw such *Umwelten* as symbolically charged, proto-cultural realms.[29] Hence, when, for example, Merleau-Ponty considered the way a crab might select an anemone as camouflage, he observed:

> The architecture of symbols that the animal brings from its side thus defines within Nature a species of preculture. The *Umwelt* is less and less orientation of a goal and more and more toward the interpretation of symbols.[30]

In modern usage, the word home is also used as an adverb, as in the phrase "to go home," as well as being used as a verb, as in the "homing instinct" of pigeons, or as a crab might be said to have "homed in" on a discarded shell. It is thus a word deeply imbued with nuances of a return, and the return to origins.

It is this same instinctual sense of home that Auden related to the significance of water,[31] evoking in his poetic reflection "In Praise of Limestone" the shared materiality of marble sculptures and the human bodies they were designed to immortalise:

> ....But if
> Sins can be forgiven, if bodies rise from the dead
> These modifications of matter into
> Innocent athletes and gesticulating fountains
> Made solely for pleasure, make a further point:
> The blessed will not care what angle they are regarded from,
> Having nothing to hide. Dear, I know nothing of
> Either, but when I try to imagine a faultless love
> Or the life to come, what I hear is the murmur
> Of underground streams, what I see is a limestone landscape.[32]

This bears a profound sense of our home residing in matter itself, and a shared ontological communality that Auden calls on in an age of war no longer shaped by the possibility of bodily resurrection. He suggests a form of love that can be nothing other than one shaped by the fault lines of our innate connections with other organisms bound by the same cycles of life and mortality. Thus, the murmur of Auden's stream is not the voice of a *River Styx* between our world and another, but the invocation of a subterranean, phylogenetic stream in the lives of all beings emerging, like ourselves, from the ocean. It is simply the desiccated, urban model of time that obscures the planetary origins of the city or the evolutionary oceanic origins of human bodies. Though as Neil Shubin has argued, we each have an "inner fish" that remains, so to speak, deep inside us, if only as an ancient biological blueprint. [33]

Like many of the marble surfaces of the urban fabric, Darwin's portrait in the Natural History Museum is comprised of the ossified ancient shells, or exoskeletons of countless millions of marine beings. And by thinking through the material histories of such human artefacts it is possible to address Helmreich's problem of an "alien ocean" disconnected from urban life: "How might marine microbes be imported into the secular, civic, or even spiritual apprehensions of the ocean world?" [34]

Because while it is true we are only able to perceive such microscopic creatures in marble or limestone long aeons after their death, the same could be said for our everyday perception of the distant stars, which though brilliant today, are suns that expired long ago. We generally sense urban stone visually in much the same way as we sense the stars: as fragments of worlds dislocated in time, yet a Darwinist reappraisal of our visual sense of the city discloses its deep connection to the ocean.

On the face of it, it would seem that reflections on deep time are oddly out of step with the urban pace suggested by the points of light on NASA's Black Marble, though in an era of rapid climatic and environmental change such assumptions must inevitably shift. There is some evidence, for example, that the decorative Cotham marble of southwest Britain used widely in Victorian buildings was formed during the End-Triassic extinction event[35] when 90 per cent of marine species died out, and marine invertebrates suffered the greatest losses. If the most prized shining stone of our cities was indeed the result of such mass marine extinctions, in the present-day context of a world on the crest of a sixth wave of extinctions unprecedented in human history, this marine history suggests further cause for reflection on the biological origins of the urban fabric. The acknowledgement of the oceanic origins of the built environment has been very slow, yet there has also been a surprisingly rapid, countervailing turn in twentieth-century film where marine mammals have entered visual culture as agents enacting human feelings about the ocean.

A distant marble

NASA's *Black Marble* series may not take creative license with the ocean and its creatures, but as a human cultural artefact it nonetheless tells its own story of anthropocentric self-imaging. In NASA's photographs and videos, the North Atlantic appears as a black, velvety expanse flanked by the sparkling lights of the Eastern seaboard of the US to the West, with the blaze of light of Western Europe to the East, which is joined by the lights of Russia in the North. In the Bering Strait, the dimmer lights of the Russian Chukchi peninsula are mirrored by those of remote Alaska, while in nocturnal images of the Arctic ocean, the *Aurora Borealis* appears as a spectacular ring of vivid green fire around the upper rim of the earth, and is mirrored by the flaring emerald light of *Aurora Australis* around Antarctica at the geomagnetic Southern pole (Figure 12.4, see plate section).

In the *Black Marble* series, such brilliant *Aurorae* are a naturally formed contrast to the illuminated cities and the dark zones between signs of human life, or the oceans between land masses. Here, rings of green light are shaped by solar activity extending to electrically-charged solar wind particles colliding with oxygen molecules in the cold ionosphere. These vast swathes of emerald light are a material sign of oxygen, and as such, are a compelling reminder of the atmospheric life of the earth; whereas in the dark zones of the satellite images, half of the world's oxygen continues to be produced invisibly by the dynamic agency of phytoplankton: the minute organisms at the crucial primary base of the oceanic food chain. Through a process of photosynthesis, phytoplankton convert carbon dioxide and regulate global climate, a process that now extends to converting excessive levels of anthropogenic CO_2 while at the same time the oceans are subject to slow suffocation from global warming and acidification.

In the full light of day, to most of us the ocean looks like a shining, pristine expanse of sparkling water because anthropogenic effects such as acidification, mercury poisoning, micro-plastic pollution or gradual warming accrue stealthily in ways essentially invisible to human vision, or even to the human imagination. Yet these processes are slowly, but surely, afflicting every ocean on earth as critical factors in what Rob Nixon aptly described as the "slow violence" of widespread environmental erosion.[36] Such biochemical and physical processes constitute one of the most acute environmental problems of our era that will affect people everywhere, yet outside science they remain little understood,[37] which is why the urban, cultural imagery of oceanic life is so critical.

When Zygmunt Bauman envisaged a condition of *Liquid Modernity*,[38] his central analogy was drawn between the flow of late modern fiscal liquidity and marine metaphors of globalization. As a model of late modernity, Bauman's view privileged perpetual human consumption in ways connected to Steve Mentz's notion of modernity as a shipwreck leading to a point of ecological exhaustion.[39] Yet liquid modernity continues to prevail within social processes of abstraction, where it is as diffuse as the steams of digital data in which we are increasingly immersed.

Though reflective star gazing is no doubt one of the oldest cultural forms, futuristic fantasies of space as a final frontier are often imbued with the same aesthetic sense of seamless progress evinced by the streamlined skyscrapers of the modern city. At the lived, street levels of the city, however, late modernity remains a labile and globally unstable process, in which to encounter a massive ocean creature is to compare human exceptionalism with a much more powerful sense of physical agency. Similarly, an oceanic poetic of deep time suggests an evolutionary *longue durée* that, just as with cosmological time, is not only permanently beyond human control or management, but has an ontological status that cannot even be fully conceived by the human mind.

As I have aimed to show, however, an aesthetics of temporal rupture and spatial dislocation offers a compellingly affective means of enabling urban culture in its regressively slow turn towards a more biotic, and less anthropocentric planetary imaginary.

Notes

1 United Nations Department of Economic and Social Affairs, Population Division, *World Urbanization Prospects: The 2014 Revision, Highlights* (New York: United Nations, 2014), https://esa.un.org/unpd/wup/publications/files/wup2014-highlights.pdf.

2 Linda Williams, "The Anthropocene and the Long 17th Century 1550–1750," in *The Cultural History of Climate Change*, eds. T. Bristow & T. Ford (London and New York: Routledge, 2016), 87–107.

3 Peter Sloterdijk, *In the World Interior of Capital: Towards a Philosophical Theory of Globalization*, trans. Wieland Hoban (Cambridge: Polity, 2014).

4 University of Arizona, "First evidence of surprising ocean warming around Galápagos corals," *ScienceDaily*, www.sciencedaily.com/releases/2018/02/180221131836.htm (accessed May 27, 2018). E. S. Poloczanska, M. T. Burrows, C. J. Brown, J. Garcia Molinos, B. S. Halpern, O. Hoegh-Guldberg, et al., "Responses of marine organisms to climate change across oceans," *Front. Mar. Sci.* 3:62 (2016). doi: 10.3389/fmars.2016.00062

5 J. Vince and B. D. Hardesty, "Plastic pollution challenges in marine and coastal environments: from local to global governance." *Restor. Ecol.* 25 (2017): 123–8; S. B. Borrelle, et al. "Opinion: why we need an international agreement on marine plastic pollution." *Proc. Natl Acad. Sci. USA* 114 (2017), 9994–7; K. Isensee and L. Valdes, *GSDR 2015 Brief: Marine Litter: Microplastics* (IOC/UNECSCO, 2015). Anna V. Ivanina and Inna M. Sokolova, "Interactive effects of metal pollution and ocean acidification on physiology of marine organisms," *Current Zoology* 61:4 (1 August 2015): 653–68. https://doi.org/10.1093/czoolo/61.4.653

6 J. C. Orr, V. J. Fabry, O. Aumont, L. Bopp, S. C. Doney, R. A. Feely, A. Gnanadesikan, et al., "Anthropogenic ocean acidification over the twenty-first century and its impact on calcifying organisms," *Nature* 437 (2005): 681–6. doi: 10.1038/nature04095; S. C. Doney, V. J. Fabry, R. A. Feely and J. A. Kleypas, "Ocean acidification: The other CO2 problem," *Annual Review of Marine Science* 1 (2009), 169–92.

7 Stuart B. Capstick, Nick F. Pigeon, Adam J. Corner, Elspeth M. Spence and Paul N. Pearson, "Public understanding in Great Britain of ocean acidification," *Nature Climate Change*, 6 (2016), 763–7; S. Gelcich, P. Buckley, J. K. Pinnegar, et al., "Public awareness, concerns, and priorities about anthropogenic impacts on marine environments," *Proceedings of the National Academy of Sciences of the United States of America.* 111:42 (2014),15042–7.

8 Clive Hamilton, "No Escaping the Blue Marble," *The Conversation*, 20 August 2015, https://theconversation.com/no-escaping-the-blue-marble-46454 (accessed January 12, 2016).

9 Michel Serres, *The Natural Contract*, trans. Elizabeth MacArthur and William Paulson (Ann Arbor, MI: University of Michigan Press, 1995); Linda Williams, "Between Hermes, Gaia and Apollo 8: Michel Serres and the Philosophy of Science as Communication," *Access: Critical Perspectives on Communication, Culture and Policy Studies* 26:2 (December 2007), 33–45.

10 "Aquatic Dead Zones," https://earthobservatory.nasa.gov/IOTD/view.php?id=44677 (accessed August 2018).

11 Stefan Helmreich, *Alien Ocean: Anthropological Voyages in Microbial Seas.* (Berkeley: University of California Press, 2009), 5.
12 Roger Fouquet and Peter Pearson, "Seven Centuries of Energy Services: The Price and Use of Light in the United Kingdom (1300–2000)," *The Energy Journal* 27:1 (2006), 155.
13 Philip Hoare, *The Sea Inside* (London: Fourth Estate, 2014), 210.
14 Jonathan Amos, "Museum's 'Dippy' dinosaur makes way for blue whale," BBC News [set BBC News in italics] 29 January 2018, http://www.bbc.com/news/science-environment-31025229 (accessed August 1, 2015).
15 Sir Michael Dixon, quoted in "Natural History Museum unveils Dippy's replacement, a giant blue whale called Hope," *The Telegraph*, 13 July 2017 https://www.telegraph.co.uk/news/2017/07/13/natural-history-museum-unveils-dippys-replacement-giant-blue/ (accessed September 19, 2018).
16 Neil Brenner and Christian Schmid, "Planetary Urbanization," in *Urban Constellations*, ed. Matthew Gandy (Berlin: Jovis, 2013), 10–13.
17 Val Plumwood, *Feminism and the Mastery of Nature* (London: Routledge, 1993).
18 Lynn Margulis, *Symbiotic Planet: A New Look at Evolution* (New York: Basic Books, 1998).
19 Linda Williams, "Reconfiguring Place: Art and the Global Imaginary," in *The SAGE Handbook of Globalization*, eds. Manfred Steger, Paul Battersby and Joseph Siracusa (London: Sage Publications, 2014), 471.
20 "This is a great day with the young people of," *The Times*, 18 April 1881, 9.
21 J. Barrie Bullen, "Alfred Waterhouse's Romanesque 'Temple of Nature': The Natural History Museum, London," *Architectural History* 49 (2006), 257–8.
22 The museum was commissioned by Sir Richard Owen, a critic of Darwin and Huxley's adversary. By 1927, an Owen sympathiser ensured that Darwin's marble portrait was moved to a less visible spot, to be replaced by a bronze portrait of Owen, and it was not until the 2009 bicentenary that the Darwin statue was returned to its original position.
23 This is in contrast to the bronze statue of Owen, a standing figure in an academic gown.
24 The other main sources were an English edition of Alexander von Humboldt's *Personal Narrative of Travels to the Equinoctial Regions of the New Continent* (first English edition, 1814–1829), the King James version of *The Bible*, and Milton's *Paradise Lost* – see Linda Williams, "Darwin and Derrida on Human and Animal Emotions: The Question of Shame as a Measure of Ontological Difference," *New Formations: A Journal of Theory/Culture/Politics* 76 (Summer 2012): 21-37 for my account of the influence of the last two on Darwin.
25 W.H. Auden, "In Praise of Limestone," in *Collected Poems* (New York: Vintage Reprint, 1991), 540.
26 So Kawaguchi A. Ishida, R. King, B. Raymond, N. Waller, A. Constable, S. Nicol, M. Wakita and A. Ishimatsu, "Risk maps for Antarctic krill under projected southern ocean acidification," *Nature Climate Change* 3 (2013): 843–7; Jean-Pierre Gattuso and Lina Hansson, *Ocean Acidification* (Oxford: Oxford University Press, 2011); Orr et al., "Anthropogenic ocean acidification."
27 The highlighting of the connections between the worlds of marine plankton and our own is not intended to suggest that ontogeny recapitulates phylogenetic processes in the literal way Haeckel understood it (Stephen Jay Gould, *Ontogeny and Phylogeny* [Cambridge, MA: The Belknap Press, 1977], 3), but rather that as I go on to argue with reference to Neil Shubin (2009), that there are nonetheless phylogenetic resemblances between creatures living today and those of the distant past that may occur in the ontogenetic scope of an individual life-span.

28 Jakob von Uexküll, *Theoretical Biology*, translated by D.L. Mackinnon (New York: Harcourt Brace, 1926), 190.
29 Louise Westling, *The Logos of the Living World: Merleau-Ponty, Animals, and Language* (New York: Fordham University Press, 2014).
30 Maurice Merleau-Ponty, *Nature: Course Notes from the Collège de France*, trans. Robert Vallier (Evanston, IL.: Northwestern University Press, 2003), 176.
31 W. H. Auden, *The Enchafèd Flood: or, The Romantic Iconography of the Sea* (New York: Vintage Books, 1967).
32 Auden, "In Praise of Limestone," 542.
33 Neil Shubin, *Your Inner Fish: The Amazing Discovery of our 375 million year old ancestor* (New York: Random House, 2009).
34 Helmreich, *Alien Ocean* 4.
35 Y. Ibarra, F. A. Corsetti, S. A. Greene, and D. J. Bottjer, D.J. "Microfacies of the Cotham Marble: A tubestone carbonate microbialite from the Upper Triassic, South Western UK," *Palaios* 29:1 (2013): 1–15.
36 Rob Nixon, *Slow Violence and the Environmentalism of the Poor* (Cambridge, MA: Harvard University Press, 2013).
37 Capstick et al., "Public understanding in Great Britain of Ocean Acidification"; Gelcich et al., "Public awareness, concerns, and priorities.
38 Zygmunt Bauman, *Liquid Modernity* (Cambridge, UK: Polity Press, 2000).
39 Steve Mentz, *Shipwreck Modernity: Ecologies of Globalization, 1550–1719* (Minneapolis: University of Minnesota Press, 2015).

13 Siren and silent song

Evolution and extinction in the submarine

Josh Wodak

1 Let me take you down...

The Beatles song "Strawberry Fields Forever" opens by inviting the listener to join a journey down to a place called Strawberry Fields.[1] Over the four times this refrain is sung, John Lennon invites the listener to reminisce in a halcyon time. The time of innocence is his childlike wonder at the world. The site of innocence is Strawberry Fields, described by adult Lennon as "an old Victorian house converted for Salvation Army orphans."[2] During his childhood this orphanage abutted a wooded garden around the corner of his home in Liverpool. "As a kid I used to go to their garden parties with my friends ... we'd all go up there and hang out and sell lemonade bottles for a penny and we always had fun at Strawberry Fields."[3] Death, via the orphanage, and life, via the adjoining garden, are springboards for lyrics about becoming, being and memory.

This essay begins as a riff on the ethos and milieu of Lennon's "Forever." It extends – onwards and downwards – toward the realm of the submarine over the half-century since the song's release. It focuses on pop, due to that mode's ability to capture the *zeitgeist* of any moment in time as well as its pervasive cultural impact across time. Dick Clark declared that "popular music is the soundtrack of our individual lives. Anything that ever happened to you, good or bad, was scored with the music you listened to."[4] Music is as much a contributor to submarine imaginaries as other artistic forms. Over the past several decades, it has provided a soundtrack to the rapidity of biophysical change at the outset of the 21st century.

In the following, I frame the *zeitgeist* of nascent environmentalism when "Forever" was released, and trace how two descendants of the Beatles – Radiohead and Beirut – extend this invitation toward a contemporary *zeitgeist*: evolution and extinction in the submarine. How, I ask, can we sense evolution and extinction through pop music? In what manners does pop music refract – and inform – popular science, and public environmental consciousness? The essay concludes with the contemporary, and beyond it – to speculate about the future of evolution and extinction, of the human and the more-than-human – as marine species disappear *en masse* in the course of the Sixth Extinction.

2 Let me take you back...

The Beatles' "Forever" appeared just as the terrestrial focus of nascent environmental awareness began to be spurred by and toward the submarine realm. *The Undersea World of Jacques Cousteau* in particular catalysed an emergent submarine sensibility as it "brought life under the seas to public attention for the very first time."[5] Limits – of life, habitability and human dominion – are central to Cousteau's *World Without Sun*, his 1964 Oscar-winning documentary about aquanaut research in underwater habitats. These aquanauts are probing life at the limits, for one week in Conshelf I in 1961, and for 30 days on Continental Shelf Station II in 1964. The underwater limits facing the aquanauts are palpable reminders of what happens when we literally and metaphorically get out of our depth.

The Beatles probed these limits in their music and filmmaking, taking inspiration from Cousteau's work. "Yellow Submarine" uncannily describes Cousteau's submarine vessels, which featured the same iridescent yellow colour scheme. For instance, *Submarine Anorep I* was built in 1966, the same year as the song. "Yellow Submarine" followed a series of recent encounters the Beatles had with aquatic ecosystems – standing on a coral reef in the Bahamas for their 1965 music video for "Another Girl," staying on board a yacht during their 1966 Philippines tour,[6] and Lennon writing "Forever" on location in the seaside Spanish city of Almeria in 1966 while acting in Richard Lester's film *How I Won the War.*[7] Ringo Starr, who sang lead on "Yellow Submarine," would later write "Octopus's Garden," inspired by a boating holiday in Sardinia in 1968. "Octopus's Garden" continued the same themes of existing in underwater habitats as "Yellow Submarine," describing sanctuary and refuge in underwater human settlements. The song uncannily describes Cousteau's Continental Shelf Station II, which explored the limits of human habitation underwater. Ostensibly for children, both songs in fact deal with proverbial refuges sought by adult selves. They conjure an invitation to come down to join an underwater refuge from a storm brewing in the atmosphere. An invitation to return to a realm of innocence, like childish wonder at the world.

Cousteau's underwater habitats presented aquatic "spaceships" for humans set against the inhospitable ocean. Occurring during the space race, depictions of aquanaut and submarine life and its limits formed palpable exemplars of the spaceship earth concept. Therein: earthly life ekes out an existence in a vessel interior, as survival is not possible on the exterior. Similarly, underwater humans can only survive on the interior of aquanaut habitats. The iconic Earthrise photograph rendered the concept of Spaceship Earth in sharp relief, by highlighting Earth as a life-supporting yellow submarine/spaceship environment encased in otherwise inhospitable surrounds.

3 The advent of the Anthropocene

Post-war environmentalism, like post-war pop music, was pervaded by naïveté. Scientific methods of researching the life aquatic were not immune: Cousteau's first feature documentary film, *The Silent World* (1956), features a sequence in which the filmmakers blow up a coral reef with dynamite to survey the species. In the scene immediately following, Cousteau justifies this by saying that it "is an act of vandalism, but the only method enabling us to list all the living species."[8] The list is given greater credence than life itself.

World Without Sun featured ground-breaking science in the then-nascent field of paleoclimatology, which pioneered the use of coral reef core samples to reconstruct past climate records. Coral produce year-on-year accretions that collectively create the reef itself, which can become a living, recording structure for thousands of years. As these accretions record coral sensitivity to ocean warming and acidification, reefs are used for paleoclimatological reconstruction. With 1/3 of marine species dependent on coral reefs for their survival, even though reefs constitute 0.1 per cent of ocean surface area[9], coral indeed lie at the heart of the life and death aquatic.

Sixty years after the release of *The Silent World*, French novelist and filmmaker Gérard Mordillat decried the irony of the title. He found it "fitting" with regard to how "today the massive destruction of coral reefs, the extermination of marine animals, hunting, pollution, the cynicism of all governments in the name of science, research and profit goes on in silence."[10] Sound and silence permeate historical and contemporary imaginaries of submarine realms: the book-jacket endorsement of marine scientist Carl Safina's *Song for the Blue Ocean: Encounters Along the World's Coasts* declares that "Safina's *Song* is the *Silent Spring* of our time."[11] The "our time"[12] is the 21st century: the *Silent Spring* analogy was added to the 2010 edition of *Song*. Safina, born in 1955, the year after *The Silent World* was released, is broadly representative of the profound changes in sensing the submarine realm between "our time"[13] and the 1967 of his 12-year-old self, when *Forever* was released and he found "the pebbly shore of Long Island Sound," near his childhood home, "unspeakably beautiful."[14]

Safina opens his *Song* with how his adult becoming is founded on ecological tension:

> When I was a boy, on warm spring evenings in the rich light before sunset my father would often take me down to the pebbly shore of Long Island Sound to hunt striped bass ...The world seemed unspeakably beautiful and – I remember this vividly – so real. It seemed so real ... When one is growing up with a sense of place, the world seems secure and filled with promise.[15]

This halcyon time is immediately followed with a lament for how this shore and its abutting aquatic ecosystem were soon after destroyed by coastal development. Safina tells the story of this disenchantment as a *Song* – a form of music that could recall both pop, as well as lyrical and childhood expression. Safina's book is a requiem for the seismic disappearances of the life aquatic in the three decades following this childhood experience. Reflecting in 1999, his reminiscence to when "the world seemed unspeakably beautiful" is Safina to his 12 year old self of 1967: the same year "Forever" was released.

Like the Earthrise photograph, *Song* suggests that the planet is dominated by marine, rather than terrestrial, realms. If we really want to go down, then it is to the marine not the terrestrial realm: "we live submerged at the bottom of an ocean of air," revealed Renaissance mathematician Evangelista Torricelli.[16] We live at the boundary between two oceans, at the bottom of the one above, and at the top of the one below. Lennon's invitation to let the listener be taken down can be to dwell at the bottom of this gaseous ocean, or to take flight to the underwater refugia of "Yellow Submarine."

Pop is the barometer of how far we have come, or fallen, since the halcyon days of naive post-war childhood environmentalism. In "our time" the uptake of Lennon's vision is the CIA naming a secret prison within Guantánamo Bay Naval Base "Strawberry Fields" because "the detainees would be held there, as the lyric put it, 'forever.'"[17] The irony becomes only more beguiling as sites of geopolitical atrocity, such as Guantánamo Bay, also manifest unintentional refugia for the life aquatic. The waters around the US Naval Base are a major site for scientific research about aquatic ecosystems due to their relative health, owing to the US enforcing stringent local activity, such as fishing, and Cuban government policy restricting tourism and development across Cuba. Due to this relative health, the marine area abutting Strawberry Fields prison is used for baseline surveys of aquatic ecosystems for the entire Caribbean and Florida Keys region. The region beyond the Naval Base waters is representative of non-baseline marine ecosystems. These are the ongoing redefinitions of the 'new normal', subject to the progressively deleterious effects Cousteau began documenting in *The Silent World* in 1954: overfishing, rapid coastal population increase, and associated industrial effluent onto reefs.

4 "Our time": of song, sound, and silent spring

In "our time"[18] another British band has taken up this mantle of providing the soundtrack *en masse* for vast swathes of a generation. Hailing from Oxford, and formed in 1985, Radiohead are widely regarded as the most influential band of their generation, courting such critical, commercial and chart-topping favour that they are known as "the Beatles of the 21st Century." Like the Beatles, Radiohead's influence transcends popular and avant-garde cultures, and similarly extends to the politics of the military-industrial complex. Edward Snowden's revelations of global government surveillance included that GCHQ named one of their surveillance programs "Karma Police" after the eponymous 1997 Radiohead song, "suggesting the spies may have been fans."[19] The pervasive cultural impact of pop music knows no bounds: recall the CIA naming their secret prison "Strawberry Fields" after the eponymous Beatles song.

However, unlike the Beatles' conviviality about their contemporaneous environmental and political context, Radiohead pursue confronting subject matter without peer in the wider terrain of pop culture:

> Radiohead have always been politically, socially and ecologically sensible, criticizing the alienation, dehumanization and the technological development of the modern world, which are the most common topics of their songs. And despite of all the success and popularity, they have still remained quite an 'underground', alternative image with no scandals, no excesses, no extravagant live shows, and not drawing attention to their private lives.[20]

The principal focuses over their more than 150 songs have been connections between biodiversity loss and climate change, induced by capitalism, consumerism, and socio-political inequality.

These connections are frequently explored with aquatic motifs of snow, ice, water, waterfalls, waves, lakes, rain, rivers, floods, the ocean, sea level rise, and encounters with marine life. The aquatic realm is also explored in relation to terrestrial human impacts, such as their song "Idioteque" (2000). Widely considered one of the first pop songs about climate change, "Idioteque" features lyrics about a coming ice age, fear of the gravity of the situation and societal myopia, which instead is focused on consumerism and the instant gratification of smartphones.[21] The song "create[s] an apocalyptic imagery which also functions as a critique of the modern society, climate change and alienation in the age of digital technology and virtual communication."[22]

Two Radiohead songs offer particularly pertinent invitations for taking us down to the sense the submarine realm in "our time"[23]: "No Surprises" and "Bloom." "No Surprises," released in the same year Safina wrote *Song,* in 1997, concerns the mundanity of suburban existence, the "great acceleration" of suburban growth, and, presciently, greenhouse gases emitted by vehicles commuting between suburbia and the city.[24] The video features only a close up of the face of Thom Yorke, lyricist, lead songwriter and band frontman, staring directly into the camera. His head is completely encased in a glass helmet, recalling the aquanaut helmets Cousteau pioneered in *World Without Sun*. Over the 3:50 song duration the helmet starts out empty then fills to capacity with water, causing Yorke to hold his breath beyond the human tolerance threshold (through a cinematographic trick of high speed filming played back at regular speed), before being emptied. Yorke sings the song lyrics with each line projected onto the helmet as he sings it.

The aquanaut symbolism is augmented with astronaut connotations. During the video's beginning and end, choreographed electronic lights flash in the glass helmet reflection, using an aesthetic that recalls the 'starman' sequence in *2001: A Space Odyssey*. In Cousteau, Kubrick and Yorke's domain, humans are attempting to probe the limits of underwater or outer-space habitats. The symbolism also recalls the planetary boundaries of spaceship earth being breached, causing its internal oxygen-based life support system to be replaced by a water-based system that is only life supporting for aquatic organisms. The 60 seconds where Yorke holds his breath because the water is above his nostrils are highly confronting. Picture a person, trapped inside a flooded helmet, staring unblinkingly into the camera. He extends an invitation to let the listener be taken down into the depths of my predicament, where what is mine is yours...

Recalling Lennon and Safina's childhood reminiscence in their respective works, a recurring motif in Radiohead's work is the problematic invocation of childlike innocence. Mark Greif argues that "beside the artificial world" (such as aquanaut and astronaut technoscience) "is an iconography in their lyrics that comes from dark children's books: swamps, rivers, animals, arks, and rowboats riding ambiguous tracks of light to the moon."[25] The "No Surprises" lyrics combined with their "musical counterpoint of chimes, strings, lullaby" form "a desperate wish for small, safe spaces."[26] In so doing, the song "promises sanctuary, a bit of quiet in which to think."[27] Tellingly, Yorke sings a song about wanting a silence that is itself silent. Their music no longer acts as a paean for the submarine realm in Safina's *Song*, or a "*Silent Spring* of our time"[28]: Yorke's face continues to sing to the viewer even while holding his breath underwater, despite his want for such silence.[29]

Fourteen years later Radiohead released their eighth studio album, *The King of Limbs*, in 2011. In the intervening years the band members increasingly focused on environmental concerns, with Yorke becoming politically active in climate change campaigns to the extent he is the pop musician most widely associated with the crisis. The first track, "Bloom,"[30] explores a dimension of the submarine realm that is the only example to date in popular music: ocean acidification. The song concerns the life and death of whales, plankton, turtles, jellyfish and an ambiguously placed human protagonist. Verse one starts from the perspective of a whale roaming the oceans with wide open mouth to collect plankton.

The whale is only able to be sustained so long as the ocean blooms. Yet the critical word "bloom" is deliberately ambiguous. Phytoplankton and algal blooms are increasing due to ocean chemistry changing due to global acidification and proximal industrial effluent. Such blooms are hazardous to marine life in general, such as whales and turtles. Zooplankton blooms are far less common, and are beneficial to those species which subsist on them, such as whales. However, a bloom is also a collective noun for a large group of jellyfish. As jellyfish have no calcifying membrane they are one of the few species potentially equipped to tolerate the deep future of acidic oceans. Their ability to do so contrasts with current projections of how acidification will render the majority of marine species extinct as the ocean is acidifying at the fastest rate in the last 300 million years.[31]

Verse two brings together the impact of human activities with planetary trajectories that have anthropogenically deviated from their deep time historical pathways. The orbital path travelled by the protagonist has now been shifted into a new and turbulent trajectory, wherein the protagonist passes giant turtles and blooms of jellyfish. The lyrics retain a deliberate air of ambiguity, though their influence has extended to scientists working on ocean acidification, such as marine ecologist Davide Di Cioccio, who used the first two verses to preface his PhD on ocean acidification.[32]

Of the innumerable contemporary bands indelibly influenced by Radiohead, Beirut is one that has gained critical and commercial clout. Zach Condon, principal songwriter and performer, formed the band in 2006 in Sante Fe, New Mexico. Six months after the release of *King of Limbs*, Beirut released *The Rip Tide*, on 30 August 2011.[33] The titular song explores a dimension of the submarine realm that is the only example to date in pop music: climate engineering. Inspired by his encounter with the submarine realm when swimming in a beach while on tour in Brazil, Condon describes how:

> a rip tide took me out pretty far – I was struggling to get back in. And as I came back in, a wave crushed me and actually punctured a hole in my eardrum. It just got me thinking: these last five years of my life, me and everyone I'm close to have all been taken by this bigger force that's mostly out of our control.[34]

This experience formed the catalyst for the song, wherein Condon searches across terrestrial and submarine realms for a place to call home. Like Lennon in "Forever," he questions whether he is, or even can be, alone. In this search the refrain recalls his near-drowning experience in Brazil, wherein he sings of being lost at sea in a rolling tide, only to find, and be found by, a rip tide, which brought him safely back to land.

As with "Forever" and "No Surprises," the full subject matter only emerges when considering the song alongside its video counterpoint. Condon himself found this to be the case as the video "brought the song somewhere that I had only been able to describe to myself, now available for others to see and feel it much more as I had in the process of writing it."[35] The "somewhere" that video director Houmam Abdallah "brought the song"[36] displays a similar prescience to the video for "No Surprises." Just as Cousteau explored biophysical human limits in underwater habitats and Yorke explored his respiratory limits in the submarine habitat of "No Surprises," "The Rip Tide" too explores the limits of human habitability in otherworldly realms.

The video features a lone empty sailboat sailing itself across the open ocean. The steady percussive rhythm forms a slow funeral march, mimicking repetitive nautical sounds of waves lapping at a sailboat or a sail flapping against a mast. Our invitation is to follow the sailboat on its journey into the abyss. There are no signs of life other than seagulls, which appear after 45 seconds, wings beating in time with the steady percussive rhythm. Over the following two minutes the number of seagulls swarming around the sailboat increases, as if they are attracted by carrion on board, as the sailboat is evidently not a fishing vessel. The boat then enters a transformative 'storm' whereupon they all leave. Throughout the remaining 1:50 this 'storm' completely reconfigures the sky, which changes from typical light sky blue to a kaleidoscope of colours, textures, patterns and forms, achieved by overlaid special effects from dropping different coloured ink into a water-filled glass tank.

The transition from typical sky blue to highly volatile and changeable colours recalls the more visual side effects of the foremost climate engineering proposal, which is to inject sulphur particles into the stratosphere to mimic the planetary cooling effects of large volcanic eruptions. The video recalls such sudden atmospheric modification and its optical side effects, which "would have turned the deep blue skies of rural areas into a Parisian-style white haze – but also have made dramatic fiery sunsets like the Krakatoa-induced one painted by Edvard Munch in *The Scream* entirely routine."[37]

The future existence of species in "The Rip Tide" and "Bloom" is ambiguous and indeterminate. Perhaps that is fittingly opened ended, given we are living in a "no-analogue state"[38] with a "no-analogue future."[39] The "bloom" in question could be for phytoplankton or zooplankton or jellyfish. The population of some species may proliferate via capitalising on human detritus. Recall phytoplankton feeding off nutrient runoff from agricultural fertiliser and seagulls as urban scavengers extraordinaire. "The Rip Tide" video location off the Cornwall coast is an area where climate change has already demonstrably benefitted certain gull species that capitalise on warming waters,[40] while in other regions climate change has already decimated other gull species by reducing available marine food.[41] "Bloom" and "The Rip Tide" invite us to ruminate on which species will be no more in light of anthropogenic climate change and which will come to inhabit future anthropogenic biomes.

5 Endless forms most beautiful

Such pop music offers a soundtrack to accompany the rapidity of biophysical change at the outset of the 21st century. Gone is the reminiscence for childhood wonder that Lennon and Safina drew from. Such nostalgia has been replaced by a yearning to make sense of our existential predicament. For refuge from modern life in "No Surprises." For facing up to climate change in "Idioteque." For communion with the more-than-human aquatic world in "Bloom." For casting adrift a path forward to a posthuman world following a near-drowning misadventure in "The Rip Tide." These yearnings are not site-specific in the manner of "Forever" or *Song*. These contemporary songs are ungrounded and footloose – cast into the elements like the sail boat in "The Rip Tide." This is the desert island soundtrack offered by the rightful heirs to The Beatles. This is the *zeitgeist* for "our times."[42]

We are nevertheless bound to recall some sense of anchorage amidst such turbulence. For Lennon and Safina such anchorage was located in specific sites and the childhood experiences they gave rise to. The site that served as an "image"[43] for Lennon's ruminations lives on every time you accept his invitation to to be taken down. However, the physical Salvation Army site is no more: knocked down in the decade following Lennon's song, replaced by a smaller home subsequently closed in 2005. Even the iconic red park entrance gates, which served as the symbolic epicentre of fans' pilgrimages, did not survive. They were replaced with replicas in 2011 – the geo-piety is now toward a simulacra. Following Lennon's own untimely death in 1980, his cultural imaginary was embodied in *Strawberry Fields Forever*, a memorial in New York City's Central Park, located around the corner from where he was murdered.

Strawberry Fields in New York City and Liverpool share something else in common, beyond surrogate memorials to Lennon and his invitation in "Forever." The Liverpool site is 2km from the Mersey, the nearest river, and 10km from the ocean at the Mersey mouth. The New York City site is 1.3km from the Hudson, the nearest river, and 8km from the ocean at southern Manhattan. The sites are respectively 77m and 25m above sea level.[44] Both are proximate to the deep future height of the ocean in Liverpool, and well underwater in New York City. By the time the sea has risen to such heights, the cities themselves would not even exist in any inhabitable form, and the tolerance thresholds of human civilisation for such biophysical change would have been breached long ago, like Cousteau or the Beatles' yellow submarine filling up with water.

Can this invitation to let me take you down, then, reframe how we see evolution and extinction play out in the littoral zones of cities founded in the early modern period? I am referring not to the persistence of naïve environmentalism wherein the Intergovernmental Panel on Climate Change consistently undermines the behaviour of non-linear system dynamics in its overly conservative modelling for sea level and temperature increase for the remainder of this century. A wealth of scientific literature, such as that of Kevin Anderson and Alice Bows[45] and James Hansen[46] lays out just how far we have already fallen through the looking glass and into the realm of runaway climate change.

So instead, I extend this invitation toward life forms that may evolve to colonise cities situated in the contemporary littoral zone. My hope is that this invitation is taken up by coral, who, like *homo economicus*, are colonists. Their ability to grow on substrates of metal, plastic, concrete and other mainstays of our times means that some deep future descendent of Strawberry Fields Montipora coral may come to make a home in "tropicalised" oceans above underwater urban detritus of the former Strawberry Fields sites of Liverpool and New York City. I think Lennon would find that a fitting finale for the notions of "Forever" in his song. Perhaps he was right all along that the contemporary difficulty of becoming someone will someday work out.

Only for the someone in "Forever" to be replaced with a "something" that may come to flourish in a posthuman world. So, to re-invoke Safina's 12-year-old sense that "the world seemed unspeakably beautiful"[47] we may need to broaden our horizons, "submerged at the bottom of an ocean of air"[48] as it were, toward a world whose beauty we cannot yet speak of. Darwin closes *On the Origin of Species* with the final sentence: "endless forms most beautiful and most wonderful have been, and are being, evolved."[49]

Notes

1 The Beatles, "Strawberry Fields Forever" (London: Parlophone, 1967).
2 John Lennon, quoted in Anthony Elliott, *The Mourning of John Lennon* (Berkeley: University of California Press, 1999), 59.
3 Ibid. 59.
4 Dick Clark, in Michael Arkush, "Q&A With Dick Clark: 'Popular Music Is the Soundtrack of Our . . . Lives,'" *Los Angeles Times*, February 7, 1994, http://articles.latimes.com/1994-02-07/entertainment/ca-20043_1_dick-clark.
5 Gary Haq and Alistair Paul, *Environmentalism Since 1945* (London: Routledge, 2013), 83.
6 "The Beatles leave the Philippines," *BeatlesBible.com*, http://www.beatlesbible.com/1966/07/05/the-beatles-leave-the-philippines/ (accessed May 1, 2016) and "The Beatles arrive in the Philippines," *BeatlesBible.com*. http://www.beatlesbible.com/1966/07/03/beatles-arrive-in-manila-philippines (accessed May 1, 2016).
7 Stephen Daniels, "Suburban pastoral: Strawberry Fields forever and Sixties memory," *Cultural Geographies* 13, no. 1 (2006): 36.
8 Jacques-Yves Cousteau, *The Silent World* (France: FSJYC Production, 1956).
9 Brian W. Bowen et al., "The Origins of Tropical Marine Biodiversity," *Trends in Ecological Evolution* 28 (2013): 359–66.
10 Henry Samuel, "Row erupts in France over famed sea explorer Jacques Cousteau's 'disgusting' abuse of marine life," *The Telegraph*, July 8, 2015, http://www.telegraph.co.uk/news/worldnews/europe/france/11726747/Row-erupts-in-France-over-famed-sea-explorer-Jacques-Cousteaus-disgusting-abuse-of-marine-life.html.
11 Richard Ellis, "The Lower Depths: *Song For The Blue Ocean: Encounters Along the World's Coasts and Beneath the Sea*s by Carl Safina," *Los Angeles Times Book Review*, January 18, 1998, http://articles.latimes.com/1998/jan/18/books/bk-9463.
12 Carl Safina, *Song for the Blue Ocean: Encounters Along the World's Coasts and Beneath the Seas* (New York: Holt Paperbacks, 1999), xii.
13 Ibid., xii.
14 Ibid., xii.
15 Ibid., xii.
16 Evangelista Torricelli, 1644, quoted in Gabrielle Walker, *An Ocean of Air: A Natural History of the Atmosphere* (London: Bloomsbury, 2010), 24.
17 David Johnston and Mark Mazetti, "Interrogation Inc.: A Window Into C.I.A.'s Embrace of Secret Jails," *New York Times*, August 12, 2009, https://www.nytimes.com/2009/08/13/world/13foggo.html.
18 Safina, *Song*, xii.
19 Ryan Gallagher, "From Radio to Porn, British Spies Track Web Users' Online Identities," *The Intercept*, September 25, 2015, https://theintercept.com/2015/09/25/gchq-radio-porn-spies-track-web-users-online-identities.
20 Nina Jukić, "Pop, Sampling and Postmodernism: The Case of Radiohead's Idioteque," *Proceedings of the 2nd International Conference of Students of Systematic Musicology*, Ghent, Belgium, November 18–20, 2009, 75.
21 Radiohead, "Idioteque" (London: Parlophone, 2000).
22 Jukić, "Pop, Sampling and Postmodernism," 79.
23 Safina, *Song*, xii.
24 Katie Sharp, "Six Eerie Ways That Radiohead Predicted the Future," *Mic*, April 9, 2015, http://mic.com/articles/114190/6-ways-radiohead-foresaw-the-future.
25 Mark Greif, "Radiohead, Or the Philosophy of Pop," in *Radiohead and Philosophy: Fitter, Happier, More Deductive*, eds. Brandon Forbes and George Reisch (Chicago: Open Court, 2009), 104.
26 Ibid., 104.

27 Ibid., 104.
28 Ellis, "The Lower Depths."
29 Radiohead, "No Surprises" (London: Parlophone, 1997).
30 Radiohead, "Bloom" (London: XL, 2011).
31 Bärbel Hönisch et al, "The Geological Record of Ocean Acidification," *Science* 335, no. 6072 (2012): 1058–63.
32 Davide Di Cioccio, "Ecology of the toxic dinoflagellate Ostreopsis cf. ovata along the coasts of the Campania region (Tyrrhenian Sea, Mediterranean Sea)," PhD Thesis, Università degli Studi di Napoli Federico II, 2014.
33 Beirut, *The Rip Tide* (New York: Pompeii Records, 2011).
34 Kristianna Smith, "Beirut: A Jet-Setter Settles Down," *NPR.org*, September 10, 2011, http://www.npr.org/2011/09/10/140318038/beirut-a-jet-setter-settles-down.
35 Laura Snapes, "Video: *Beirut* 'The Rip Tide'," *Pitchfork*, June 22, 2012, http://pitchfork.com/news/46943-video-beirut-the-rip-tide.
36 Beirut, *The Rip Tide*.
37 Bronislaw Szerszynski, "Colouring Climates: Imagining a Geoengineered World," in *Routledge Handbook of the Environmental Humanities*, eds. Ursula Heise, Jon Christensen, and Michelle Niemann (London: Routledge, 2017), 82–90.
38 Paul Crutzen and Will Steffen, "How Long Have We Been In the Anthropocene Era? An Editorial Comment," *Climatic Change* 61 (2003): 253.
39 John W. Williams et al., "Model Systems for a No-Analog Future: Species Associations and Climates during the Last Deglaciation," *Annals of the New York Academy of Sciences* 1297 (2013): 29–43.
40 "Seagulls feel the benefits of climate change," *Phys.org*, July 5, 2012, http://phys.org/news/2012-07-seagulls-benefits-climate.html.
41 Louise Blight, Mark Drever, and Peter Arcese, "A century of change in Glaucous-winged Gull (Larus glaucescens) populations in a dynamic coastal environment," *The Condor* 117, no. 1 (2015): 108–120.
42 Safina, *Song*, xii.
43 Lennon, quoted in Daniels, "Suburban Pastoral," 39.
44 Strawberry Fields, Liverpool, http://www.streetlist.co.uk/address/Beaconsfield-Road-Liverpool-L25-6EJ (accessed December 7, 2015), and Strawberry Fields, New York City, http://www.bikeforums.net/northeast/303521-manhattan-nyc-elevation-meters-above-sea-level.html (accessed December 7, 2015).
45 Kevin Anderson and Alice Bows, "Beyond 'dangerous' Climate Change: Emission Scenarios for a New World," *Philosophical Transactions of the Royal Society of London A: Mathematical, Physical and Engineering Sciences* 369 (2010): 20–44; and Kevin Anderson and Alice Bows, "Reframing the climate change challenge in light of post-2000 emission trends," *Philosophical Transactions of the Royal Society of London A: Mathematical, Physical and Engineering Sciences* 366 (2008): 3863–82.
46 James Hansen, "Climate Urgency," *Simulation & Gaming* 44 (2013): 232–43.
47 Safina, *Song*, xii.
48 Torricelli, in Walker, *An Ocean of Air*, 24.
49 Charles Darwin, *On the Origin of Species by Means of Natural Selection, Or, The Preservation of Favoured Races in the Struggle for Life* (London: J. Murray, 1859), 459.

14 The ocean hospital – a walk around the ward

Janet Laurence and Prudence Gibson

Australian artist Janet Laurence's installation *Deep Breathing: Resuscitation for the Reef* (2016) presents as a ward hospital, a site of care for a sick ocean reef. Here, in these symbolic wards, it is possible to participate in the diagnosis and planned treatment of the Great Barrier Reef, and to ponder the prognosis. Staying present in such environments requires courage and care. This chapter is a metaphorical walk through the hospital wards of the Great Barrier Reef, as imagined by Janet Laurence. Each image, and each corresponding textual fragment, functions as a discrete room for medical examination, and for philosophical and environmental reflection premised upon the familiar constructs of aesthetic allure.

Janet Laurence is an internationally-esteemed artist whose work is a radical intervention into ecological demise and, as an activist, her work permeates the public imaginary. *Deep Breathing* was first exhibited at the 2015 *Artists 4 Paris Climate talks*, Muséum national d'histoire naturelle, Paris. Then the Australian Museum in Sydney embraced her research and invited her to visit Lizard Island to prepare for a Sydney version of the artwork, installed in the museum foyer. The artwork is a meditation on damage wrought upon the Great Barrier Reef. This art hospital is a conceptual healing of the sick, but also recalls the experience of a hospice for the dying coral and wider ecology. Using ocean detritus, scientific specimens from museum collections, and biotic matter relating to the undersea, Laurence creates an art installation that tests the hypothesis that aesthetic encounters can disrupt our climate-change-blindness.

Deep Breathing presents nature in forms that are a diversion from the kinds of aesthetic experiences we have known before. Rather than attempting mimicry or a mirroring of what we think we see when we are immersed in a natural environment, Laurence creates a parallel world. It becomes an amalgamation of elements, re-constructed in a mode that is still familiar to us (the museum display) but dances between representation (such as conventional painting – or in this case, piles of paint pigment) and presentation (such as the equally conventional museum-style cabinet of curiosity).[1] This new aesthetic form, where science and art dovetail, has catapulted Laurence's work to the corners of the globe – from the recent exhibition *Fragil* (Ecuador, 2017) to *The Underlying* (October Salon in Belgrade, 2017) to Paris *Climate Talks* 2016 and her 2017 work *The Matter of the Masters*, exhibited at the Art Gallery of New South Wales, Sydney.

The experience of Laurence's work is political (environmental), immersive (physical), and affective (emotional). As all transformative art does, *Deep Breathing* first arrests our curiosity and draws us towards the specifics of its multiple elements: glass and scientific equipment and specimens. In other words, the parts attract us. The total sum of the installation takes longer to infiltrate our intellectual imagination. Audiences can experience the sensory information of the work from all sides or as a cross-cut section of its entirety. The way the glass exhibition boxes are ordered and stacked contributes to this sense of a geological cut-through or slice of time and space. The use of mirrors on each section's base creates a sense of infinity below and above. This then creates a never-ending mesh of information and experience in all directions. Secondly, once the viewer has time to absorb the data of the presentation, its political effects settle in the viewer's cognition, probing both fears and hopes for the future of the Reef.

In *Deep Breathing*, the artist not only retrieves the spoils of the sea from natural history specimen collections, but also recreates other specimens through her alchemical artistic processes of casting and re-forming. This combination of real and artificial objects is a direct reference to the tradition of the *wunderkammer*, the cabinet of curiosity, where the strange and beautiful souvenirs from expeditions of colonial discovery were displayed in cabinets and even entire rooms of Victorian homes.[2] This combination of the artificial and the natural reflects our human relationship with nature which has become hybridised and hyper-evolved (the natural adapting to the unnatural and vice versa). It is relevant in an epoch of climate change because all species are struggling to change themselves, become something new, for survival in difficult ecological conditions.

The main concept or focus of *Deep Breathing*, however, is its role as a hospital. It exists as a place where triage, diagnosis and pathology procedures are undertaken for the "spoils of the sea" before a schedule of care can begin. Laurence's patients are the deep-sea sponges, the turtles, the myriad fishes, the cephalopods, the shellfish and the coral. These creatures require care, having been made sick by human activity. There is an undercurrent of irony in all the elements of these deep-sea works which begs the question: If humans have caused the reef's ill-health and its inhabitants' sickness, can we now cure them? In addition, do we have the moral right to appropriate that role? Can the natural environment survive and revive without human interference, without human existence?

The notion of nature as endless bounty is no longer a popular maxim with reef scientists, who are also working to better understand what options there are for a cure.[3] The Great Barrier Reef research stations are working to find out what corals are doing and to introduce new species of corals that can live in warmer water.[4] The Reef, as a world heritage site and focus of interest, is also of critical importance due to the necessary structure it provides to protect the coastline from cyclonic damage. The violent waves that have always crashed onto the sturdy reef now crash onto weak and brittle coral. Once the coral dies, it crumbles away. Dead coral is no longer porous and flexible because there are no living creatures to keep it flexible and calcified. Laurence says,

> My vision of healing is based in a "nature imaginary" but it is also based on experiments of scientists at the Lizard Island Research Centre off Queensland. There, reef elements are studied for calcification, how fishes behave, how coral survives in warmer water, when and how it dies, why there are algal blooms and why there is an explosion of Crown-of-thorns starfish that kills coral.[5]

An *aesthetics of care* is an attitude of concern and worry, adopted here in the face of the ensuing Research Centre information. The phrase was coined by SymbioticA bio-art lab, University of Western Australia, in 2002. It originally referred to the social implications of using medical technologies for artistic purposes.[6] The cultural and social effects of manipulated and damaged living systems are fraught. There are new evolutionary discoveries that are being recorded but at a snail's pace, compared to the register of threatened and endangered species, the Environment Protection and Biodiversity Conservation Act list.[7] Who will stand firm and take care, when an Adani mine development looms over the Reef and no government policy will be implemented to protect the undersea nature rights?[8] Laurence says,

> the experience and the image that the work creates is one of care but how do you get there, to a place of care? I have to express the work at a high level of empathy or engagement with the subject that creates a sense of pathos. As visual art, it mimics or imitates how we care for humans in hospital. We automatically understand that human angle of care which helps people imagine the activity of care needed to be undertaken for the reef to survive.[9]

In the end, *who* cares? And who *cares*? These two questions mark the distinction between who is bothered to be concerned, and who is prepared to put time and effort into restorative treatment. As John Clare wrote in his poem "The Yellow-hammer's Nest" (c.1820), we are leaving behind a "ruined nest."[10] Laurence recreates the nest, to remind us of the species that are ruined in the wild, whilst breathing the faintest hope that the nest might become active once more. The Great Barrier Reef is bleaching so quickly, it is hard to witness without wanting to shield our eyes from the glare. Its acceleration into extinction is irrefutable, and so the multiplicity and diversity that we see in Laurence's mirrors and reflections are a moral compass, a means of drawing our attention closer to what has happened.

Of course, this is exactly the point that Laurence is making. She has been an advocate for animals, plants and at-risk ecologies for many years, and her work has consistently drawn our attention to uncomfortable truths. Her installations often include clutches of seaweed, sponges, varieties of coral, natural paint smeared on glass, mounds of biotic ash or samples of dirt from a compromised ecology. In *Deep Breathing*, medical tubing is threaded between the varying scientific glass beakers and flasks, while projections of film in the background amplify the sense of deep water. This creates fluidity between the specimens (nature) and the scientific materials (culture) and the endless sea (infinity). The hospital includes alternative homeopathy which Laurence explains here: "I dived down through the water to the coral and attached healthy, coloured coral to the sick bleached coral for homeopathic care. It was a symbolic homeopathic gesture."[11] This is an instance of how the artist attends the sick and dying.

Wards 1–12 are presented in colour in Figures 14.1–12 of the plate section.

Figures 14.1–12: Janet Laurence, *Deep Breathing: Resuscitation for the Reef*, 2016, mixed media: scientific glass, cast resin, silicon tubing, silk thread, pigments, fluids, marine specimens. Courtesy of the Australian Museum, Sydney, and the Muséum national d'Histoire naturelle, Paris. Photographs by Benjamin Huie and Lena Galangau-Querat.

Ward 1. The obscenity of the coral collapse

In this first ward, the tiny medical glass vials, approximately two centimetres long, perch on the tip of each coral finger. There is an obscenity about this deep-sea hybrid being, which functions as metaphor for human abuse of sea life. Obscene, because the coral is like a drunkard sucking on a bottle, or a morphine addict receiving yet another hit. The white coral, tinged with a gangrenous green, is not looking in good health. Even the flask with green matter suggests substance abuse, a concoction heated up in a meth lab.

Both the human and nonhuman are implicated: sea life has become part of anthropocentric life, rather than the reverse. This is the posthuman condition: in this case, the emergence of a kind of life outside the conventional notions of species. It is a problematised life that humans cannot grasp in any moment of linear chronological time.[12] Human and nonhuman have together turned our deep-sea life into crippled substance addicts, through abuse and adaptive evolution. With all

Figure 14.1 The obscenity of the coral collapse

our imperious hubris and all our imperial sanctimony, humans have compromised consecutive ecologies with acts of discovery, colonial exploration, and resource acquisition. The vials appear to be restoring the un-restorable, but the cadaver is not responding to treatment because the treatment is further poison. This is the obscenity the reef now speaks.

White coral represents purity and other worlds. Many a middle-class Australian family home sports a large piece of designer coral on its coffee table. Such a piece has become a home-design item, almost a cliché and almost always a copy. What is this domestic coral display if not a dream of submarine life, a sense of watery relief from living on our mostly arid continent? Obscene ghosted coral. The green vials attached to this pure white coral accentuate the punctum of impurity.

Ward 2. The warp of water, the geology below

Much of Janet Laurence's installation work comprises reflection, mirroring and viewing through glass, through concave and convex lenses, through medical flasks, and through liquid. As part of a ward hospital, this refers to the observing and monitoring of the sick. However, there is a kind of purposeful obfuscation or warping of the view at work too. This obfuscation suggests the picturesque tradition: a view that is carefully obstructed (for example, by an old ruin, a fallen log or a copse of trees) to create further mystery, allure and curiosity. The cabinet of curiosity or *wunderkammer* presentation of Laurence's work relies upon this mystery. Flasks with exotic fluids, jars with flammable formaldehyde, bizarre and curious specimens collected from barely inhabited places during colonial times: these are a cornucopia of objects the viewer strains to see. This is the "enduring

Figure 14.2 The warp of water, the geology below

interest of the naïve observer" where we are fascinated by what we can't see and we are in thrall of the scale and multiplicity of the artist's elements.[13]

Laurence plays with these conventional colonial problematics by making her own specimens too. Some objects are moulded and cast, mimicked and copied. This doppelganger effect encourages the fictional, the speculative and the futuristic character of her work. It also escalates its political nature. These artfully constructed installations of museum and artist-made specimens are a political act. Her position is an articulation of environmental advocacy, not just of outrage at government inaction but deep sorrow that our ecologies are suffering at the hands of humankind.

This ward particularly involves examining the porosity of coral. The deep undersea is a realm that incorporates the air above and the sandy base below. As such, there is a geological aspect to the reef that is sometimes forgotten, that we can't see because of the movement and darkness of deep water. After all, the coral itself is an extension of the earth's crust, but its strength is diminished. Most of us only see the uppermost tip. The brittle and bleached carcass that we now see as the Great Barrier Reef is the skeleton leftovers of a once-vibrant independent living being.

The coral, on its last legs, is still in a state of change and becoming. If Donna Haraway thinks the Anthropocene is a "boundary event not an epoch" then she could well be looking at our reef as she speaks.[14] In a similar vein, where Rosi Braidotti speaks of "becoming-earth,"[15] we are here speaking of the reef as the *becoming-sea*. The matter of matter is the reef, the live animals, the chalky substance, the fragile colour scheme – these are all elements that exist with a materiality, to which Laurence responds with warning and care.

Ward 3. The bloody thread

The red thread in ward 3 winds around a small piece of white coral. It suggests a blood transfusion being undertaken. This could be a lifeline from human to coral.

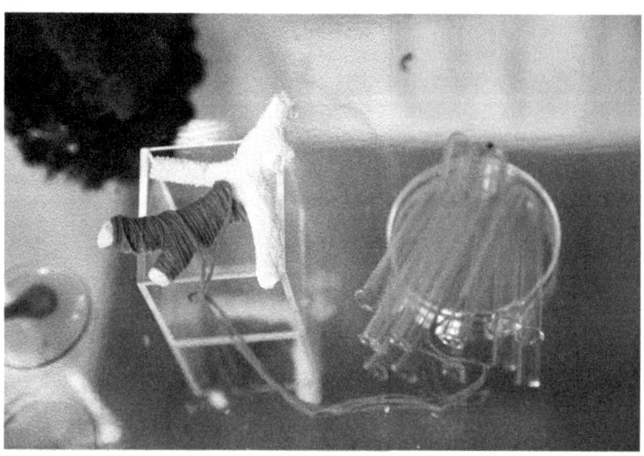

Figure 14.3 The bloody thread

The blood of humans and beasts, birds and insects, flows through to the ravaged reef coral. But this might not necessarily be a positive transfusion. It could be a poisoning, or a runoff of agriculture fertilisers into the sea. The blood of the ocean's creatures is here flowing backwards too, corrupted, terminally sick. This is the blood of fish and turtles, whales and dolphins – caught in plastic or destined to swim in seas that are too warm. It trickles back through the medical tubing in the opposite direction.

This movement of red in both directions is the last gasp for humans. If the reef is dead, the butterfly effect (with local rainforest ecologies to suffer as the next step) could be catastrophic. We can understand the movement of hospital fluids in a human realm. Once the fluids flow in the wrong direction in a hospice environment within a human body, only death awaits. This ward is highlighting the fact that the same is occurring for the reef. Laurence says, "I am also paralleling the movement of blood with the way the colour of the coral leaves the structure – my work visualises the damage and suggests the possibility of blood returning, of a restorative process."[16]

The world of human productivity has toxic by-products, the result of excessive fertilizers and agricultural pesticides, sliding downstream from farms and industry, and into the ocean foreshores. The link between human and nature is consequently toxic and red. "When I think about those pesticides, I also think of the poor shellfish that consume the toxicity, and are then weakened and prey to jellyfish etc." In Laurence's work, the pieces of coral have been triaged, wrapped in red thread, but it's only a temporary solution. The reef is poisoned. The thread also pools in the petri dishes. This is a cure, but without effect.

The thread, finally, represents the tenuousness between life and death. The slippery connections between consciousness and unconsciousness are like a vascular system – the movement is the reason for being. They are constantly moving and constantly changing, delivering nutrients and taking away waste. The problem arises when those roles become confused. Nutrients confused with waste, consciousness confused with unconsciousness. This is not just Braidotti's inhuman, it is the inhumane, the paradoxical language we use for immoral acts upon the weak and the unfortunate. Death, violence, cruelty, for Braidotti, are the artefacts of life before and after human existence. This can be read as the inhuman treatment of the reef, the inhumane.[17]

Ward 4. Swelling lament

Flask. Tube. Coral. A beaker with swollen white specimen. The white beaker specimen in question is a powder calcification, a residue that has become swollen by its aqueous encounter. Water makes everything swell. It creates oedema. Ghoulish as it may be, a dead body left in water will swell to twice its original size. Fluid retention in human bodies occurs when there is extreme heat or when a pregnant mother carries extra weight. Watery oedema: engorged and puffy, distended and bloated.

Figure 14.4 Swelling lament

The bloating is a sensory posture that can appear unrecognisable or at least unfamiliar. Unlike Alphonso Lingis's theories of human recognition and phenomenological sensibility, the swollen sea specimen is puffy and unknowable.[18] There can be no passive or active registering of what these swollen things are, only a sense of isolation and fear. The white object in Laurence's beaker has become a fictive anemone, or a product of marine imagination. None of the medical staff on the ward know what to do.

Water bloats and it also transports. Nonhuman forms have conventionally been considered philosophically inferior beings insofar as they are immobile and non-sentient, incapable of intellectual thought and unable to progress or move outside the domestic home or immediate location.[19] This bloating and swelling possess a slightly unattractive, nonhuman element, but also a kind of flowering or efflorescence. Luce Irigaray speaks of the efflorescence of the flower[20]; Laurence's white swollen specimens in the beaker are an efflorescence too, a dissolving of the inner chemical makeup of any given object to create a new external form. This kind of efflorescence is a return to knowledge, a return to the bounty and wonder of the nonhuman world. It is important to remember that, in the context of Laurence's work, they are also in a state of non-life. This is an in-between phase, sick and not cured, alive but not active, dying but not dead.

Ward 5. Conduits of colour

Janet Laurence creates an active transfusion of colour. The tiny reserves in the small glass vials may be the only remaining substances preserved. Small coral pieces are placed on the vial tops as stoppers, a playful identity kit, to create empathy between human and nonhuman. These little vessels serve as electrical plugs, receiving electricity and conducting it at the same time.

The artist is a healer, a doctor. In this ward, Laurence uses her medicinal skills to transform metal filings into inky liquids, pigments into blood that shifts across blue,

Figure 14.5 Conduits of colour

red, yellow and black. There are mixed connotations of the healer. The artist/healer/doctor is sometimes maligned, an outcast and a subject of superstition. However, the healing continues, and the elements still create change in the human, the animal, the reef. We find it hard to trust the female artist, but should we not be turning our contempt towards those who lack courage to protect the reef?

The sea has a history of untrustworthy women, capable of great power. For instance, the sirens' calls have lured men from their ships and into the arms of death. The siren, the healer, the artist. The artist makes tinctures, she makes vials of magical potions, she sees the world beyond the material and into the spiritual realm. The sea is the mystical zone, within which the artist works. Isabel Stengers writes of these maligned healer figures in cultural life who are compromised by a lack of trust, just as many are suspicious of the concept of climate change.[21] It is the women, the healers, the doctors who must be trusted. The mistake in the past has been to mistrust those with skills and knowledge, by those in positions of authority. Laurence's magic tinctures are intended to heal; she is reviving the coral by raising awareness for it.

Ward 6. Blood/chlorophyll

What role does the colour green play in the reef, as seaweed, algae, seagrasses? Green is conventionally associated with positives – green for go.[22] Blood and chlorophyll are the vascular tissues of humans and plants; the artist is connecting these two fundamental systems. She says, "these seagrasses are blooms of colour, they are complementary colours, opposites on the wheel. The red seagrass is artificial, a warning that something is not right with the reef. The seagrasses are symbolic of blood and chlorophyll, those elements that give life and light to animal and plant life." The deeper you dive into the sea, the greater the change in the colour spectrum. The decrease in light changes the coral colour to deeper blues. With depth come changes in coral colour.

Figure 14.6 Blood/chlorophyll

The sea grass blooms. Green, as the vegetal, is the biotic matter that stretches across the ocean floors, swims along the currents, and catches on reefs and rocks. Green is the colour we know deeply, and in a fundamentally psychological way. These green and natural things affect our mood, our primal urges to strive, to provide, to care. "Green" affects our sense of wonder in the face of life. The fear of losing contact with the natural world is solastalgia, a neologism referring to climate change-induced stress.[23] How we can build a positive perception of *nature* as a construct of imagination or a material and sensual collective memory is a chronic problem in contemporary society.[24] It is due to the loss of green from our everyday habitats that greenness is more than a hue of nostalgia, it has also become a keen political weapon, an activated and dynamic arena of civic and civil debate. Thinking green is the same as thinking for *the long now*, where plans and concepts need to have longevity and long-distance strategic clout.

Ward 7. The paediatric ward

These jars of fish are museum specimens from the archives. Massive losses of marine species have been caused by polluted water and lack of appropriate coral food – causing widespread infant death. In this ward, we get close to the fish, for they are the closest we get to a connection with active reef life. This is partly because of the recognisability of their eyes. Fishes in jars comprise an aesthetic we all know and are familiar with: the museum specimen in a spirit

Figure 14.7 The paediatric ward

jar. The glass of the jars and their spirit fluids amplify our view of what is inside, like a myopic lens. Laurence says, "The jar makes them both visible and invisible, the multitude we see here in these jars are no longer swimming in the ocean. Perhaps they no longer exist."[25]

Extinction threatens, non-life looms and the reflection of human survival, in the status of the reef's illness, has become clear. As Monica Bakke says "this is the unbearable closeness of being."[26] We have our noses up against the jars, to see. But the life is already gone. The jars act as bubbles, or *Umwelten* (individual environments that intersect). The liquid, like the glass of the jar, obscures and enhances our view of the fish. The surfaces are also reflective. These jars of fishes represent the extinct and the endangered, the threatened and the lost. Here they are kept as reminders of what might be revived.

Ward 8. Lost habitats

These shelves of shells are arranged as arche-fossils, objects from the past as well as from the future, as they exist outside human time.[27] Laurence says:

> For me, the shelf of shells reminds me how vulnerable they are due to the acidification of ocean. They won't grow. They won't become strong, nor will they survive the medusa jellyfish that will eat them. So, these sculptural insertions are not ordered and placed according to taxonomies, as a curator might, but as an artist does, to tell an evolutionary story. I order things to think about relationships between different specimens. I imagine how they might form together. The baby nautilus shell is part of the baby nautilus ward. These are indeterminate relationships but they create empathy through their aesthetic qualities.

Figure 14.8 Lost habitats

Shells are minutiae homes for underwater creatures.[28] Increasing acidification of water makes shells more and more fragile. Shells soften from lack of calcification and jellyfish can "take" baby shells with little difficulty. Crabs, oysters, clams, squid and octopus. The relationship between creature and shell home is intimate but the dwellings are cast aside after use, or appropriated by stronger species. But these shells are not strong, not resilient. The fluidity of shell homes contrasts to the human condition, where every human must own their own home, their own plot of land, to keep in perpetuity. This, Michel Serres argues, is the mistake of the human race. We are, in fact, no more than renters on this planet.[29] We own nothing and should think of our homes as shells. The shells drift and are adrift. They are both lost and in motion. Mobile homes. Impermanent, just like the ward of a hospital.

Ward 9. Spheres: the isolation ward

In Laurence's words,

> This is brain coral, which I cast and turned into a tentacular creature. The purple thread tentacles are teeming out of it. In this one specimen is the memory of a whole sea. When algae leave their coral, they leave each empty hole but here in my artwork it spills out from underneath – a wellspring of colour. It is a creature but it is also becoming something else. Something strangely malformed. The round dome of glass around it plays with the geometry of roundness, a repetition of form. The brain coral is in an isolation ward of a hospital. The ward is protecting it and trying to attract colour back in.

Figure 14.9 Spheres: the isolation ward

The concepts of iterative change, repetition, and becoming have been spoken about in environmental philosophy. The concept of becoming was developed by Karen Barad as a constant of possibility.[30] In a similar vein, Rosi Braidotti refers to the state of constant geo-change[31] as "becoming-earth" which leaves an open space to think of Laurence's work as a "becoming-sea," a widening circle of iterative change, not necessarily for the better.

The three-dimensional version of the circle, the sphere, is an interesting geometrical form to create and iterate in an ocean environment. Round islands push up above the sea. Islands differ from major continental land-masses in terms of Western and non-Western cultural and mythological knowledge: small islands are often named after women, visualised as women, and revered as women. Fertile mother-worlds – the roundness of the womb compares to the roundness of a small island. Underworld, cave, egg: these are spherical forms that Sloterdijk explains as "the intimacy of spheres."[32]

Is the brain coral serving as a stopper, a cave-like door, a tomb for our lost biodiversity? This burgeoning and bountiful life that Laurence expresses is also to be protected. Her entire *Deep Breathing* can be understood as a burial place, as well as a hospital.

Laurence is protecting the round brain coral, a visual statement of endless curves and the continuous, gyre-like fluidity of the deep. The motion of circular forms lulls us into a sense of harmony. In art, the circular movement of paint can hypnotise us. These round glass flask forms cause a similar effect.

Ward 10. Geometry and geology

The red fan coral is a variety of sand coral. This ward doubles as a geology lesson. "The red fan coral is amplified and magnified by and in its fluid. There is sheer

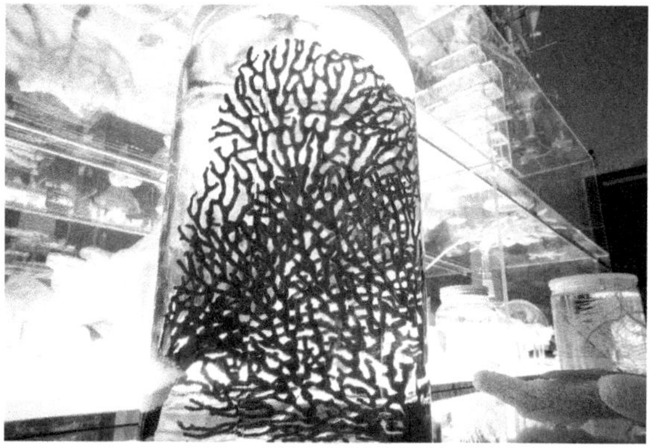

Figure 14.10 Geometry and geology

beauty in the visualising of the extraordinary geometry that these corals construct," the artist says. Its filaments are a Timothy Morton-like mesh of being.[33] This mesh is an endlessness of relations between all things, including concepts. Climate change is a concept so vast we cannot grasp it. The mesh is a means of understanding that even the hyperobject, such as the plastic lid of the jar that stores the red coral, represents an endlessness of connections and relationships that cannot be reduced. It is impossible for a plastic lid to disintegrate or break down and return to any given ecology. The environment now is made up of objects that will not return to dust but will form the latest layer or geological crust of the earth's surface. As Monica Bakke explains, geological time has great relevance at the intersection of the reef and art: "Geological time however is relevant not only when key scientific questions are considered concerning the planetary past, such as the chemical origins of life, but also in respect to the planet's future."[34]

It is hard for humans to make sense of the world outside a habit of human construction. This red coral has its own structure, its own mode of growth and geological pattern-forming. Whilst we may discuss its architecture, the coral grows despite humans, yet it dies because of humans. "If we could hear . . . the call of those who are slipping out of life forever. There we might encounter a narrative emerging from extinctions, a level of blood that connects us. . ."[35]

Ward 11. Intensive care

This image of the whole reef hospital shows how it is activated as a combined unit. Most elements are selected from the museum collection store, and each specimen is given a post-life "existence," once it has been tagged and labelled. The specimens are individual but they are also part of a collective, where their value is made clearer. As Laurence explains,

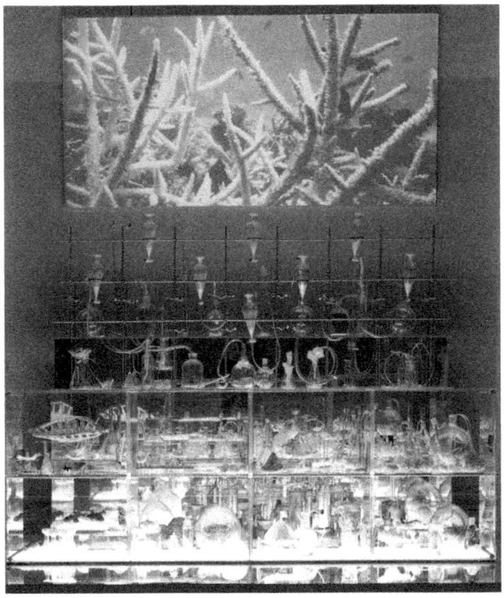

Figure 14.11 Intensive care

each specimen functions in its own singular way, it is performative as parts and as a whole. The objects belong to a scientific realm but they are altered by the active experiment of research and aesthetics. Each object has an inter-species relationship with the object next to it. They have been arranged as family stories.

The leaked-out colour is explained by Laurence:

> Elements that sit in pools of their own colour represent the last vestiges of its leaked-out colour. They're like small flowers. Colour? I ask in my work: how does it leave or fade? How can we imagine how it disappears, how it travels ... of course algae leave coral and that entails a fascinating physiology ... but the algae are the ones with the colour and we want to know how they wither up and die or shoot outwards. Coral has always been associated with colour, so the vanishing of colour is an important aspect of my work.[36]

Ward 12. A turtle's view: emergency

There are smears of paint on the glass surrounding the turtle image. These smears are reminiscent of algae and currents of different-coloured sea-water. Within this watery world sits the snugly wrapped turtle. The sick turtle was photographed by Janet Laurence at the Townsville turtle hospital. The turtle

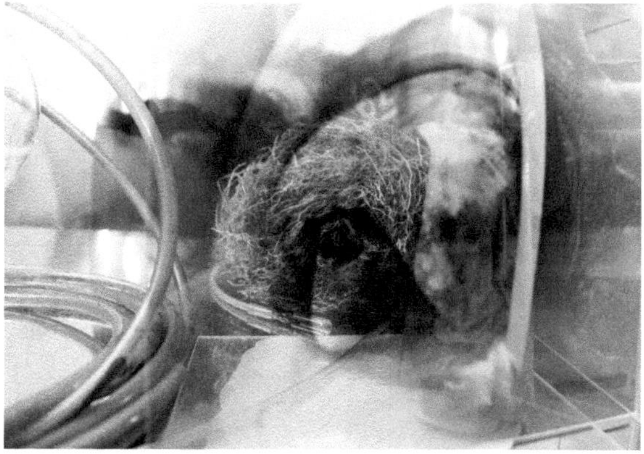

Figure 14.12 A turtle's view: emergency

ward is where sick turtles are bandaged but also where they are operated on to remove fishing hooks and plastic bags from their stomachs. X-rays are taken and care given.

The concept of care and the symbolic associations of water brings to mind Luce Irigaray's fictional correspondence with the philosopher Friedrich Nietzsche, where she argues that Nietzsche's text gives no habitat for women. His concept of "woman" is conceived as a capricious, free-flowing entity. Irigaray suggests that everything is dry, hard and arid in Nietzsche's world, where there is nothing that moves in the water, unless it is on a boat, or a man, ankle-deep. "Why this persistent wish for legs, or wings? And never gills?" asks Irigaray.[37] Laurence redresses the aridness or Nietzsche's world by submerging her artist's view below the sea.

The artist has a strong association with water as care-giver and hospital site, rather than a place of capriciousness and lack of reliability. However she also understands it is a place of danger for species. She said,

> I visited the turtle hospital in Townsville and it was disturbing to see so many turtles with their bellies filled with fishing hooks and plastic entwined around organs. There were so many of them damaged, from ghost nets and fishing collateral. Here in the sanctuary they are being cared for and looked after, yet in the wild, in their natural habitat of the ocean, we don't have the capacity to care for them. Many are lost on the long journey they take, arriving back to habitats along the shores that are no longer a source of succour or safety. The numbers of these great ancestors of the sea are decreasing. Once, so I heard, they were in such plenty you could step across parts of the harbour on their shell backs. To swim with these peaceful giants is the greatest pleasure.

End of hospital rounds

Laurence has had a long art career spanning thirty years, and the science disciplines are familiar with her work, too. She has undertaken many collaborations with plant biologists, environmentalists, landscape designers and bush regeneration experts in Australia, the UK and Germany. Her knowledge of the natural world has been developed after doing research, gathering film and photography footage in various locations such as Singapore, China, Japan and Germany. Laurence has an established role as an environmental advocate and her artworks are a cry for better care and for a quicker response time for the Reef and other at-risk ecologies. This reef hospital is Laurence's memorial. The work is composed as a lament.

Notes

1 Prudence Gibson, *The Rapture of Death* (Sydney: Boccalatte Books, 2010), 129–35.
2 Ibid., 129.
3 Bronwyn Wake, "Snapshot: Snow White Coral," *Nature Climate Change* 6 (2016): 439.
4 Lizard Island Resort, "Lizard Island Research Station," https://www.lizardisland.com.au/about/research-station (accessed September 2018).
5 Janet Laurence, interview by Prudence Gibson, 28 December 2017.
6 Oron Catts, ed., *The Aesthetics of Care?: The Artistic, Social and Scientific Implications of the Use of Biological/Medical Technologies for Artistic Purposes* (Nedlands, W.A.: SymbioticA, 2002). http://www.tca.uwa.edu.au/publication/THE_AESTHETICS_OF_CARE.pdf.
7 "Species Profile and Threats Database," Australian Government Department of the Environment and Energy. http://www.environment.gov.au/cgi-bin/sprat/public/publicthreatenedlist.pl?wanted=flora (accessed August 6, 2018).
8 Jane Gleeson White, "It's Only Natural: The Push to Give Rivers, Mountains and Forests Legal Rights," *The Guardian*, 1 April 2018, https://www.theguardian.com/australia-news/2018/apr/01/its-only-natural-the-push-to-give-rivers-mountains-and-forests-legal-rights.
9 Laurence, interview by Prudence Gibson.
10 Prudence Gibson, *Janet Laurence: The Pharmacy of Plants* (Sydney: New South Publishing, 2015), 29.
11 Laurence, interview by Prudence Gibson.
12 Rosi Braidotti, *The Posthuman* (Cambridge: Polity, 2013).
13 Gaston Bachelard, *The Poetics of Space* (Boston: Beacon Press, 1994), 107.
14 Donna Haraway, "Anthropocene, Capitalocene, Plantationocene, Chthulucene: Making Kin," *Environmental Humanities* 6 (2015): 106.
15 Braidotti, *The Posthuman*, 81–2.
16 Laurence, interview by Prudence Gibson.
17 Braidotti, *The Posthuman*, 108.
18 Alphonso Lingis, *Sensations: Intelligibility in Sensibility* (New York: Humanity Books, 1996), 55.
19 Prudence Gibson and Monica Gagliano, "The Feminist Plant: Changing Relations with the Water Lily," *Ethics and the Environment* 22, no. 2 (2017): 125–46.
20 Luce Irigaray and Michael Marder, *Through Vegetal Being* (New York: Columbia University Press), 2016.
21 Isabelle Stengers, "The Doctor and the Charlatan," *Cultural Studies Review* 9, no. 2 (2013): 11–36.

22 Prudence Gibson, "The Colour Green," in *Covert Plants*, eds. Gibson and Baylee Brits (San Diego: Punctum Books, 2018), n.p.
23 Ibid.
24 Glenn Albrecht et al., "Solastalgia: The Distress Caused by Environmental Change," *Australasian Psychiatry* 15 (2007): S95–S98.
25 Laurence, interview by Prudence Gibson.
26 Monica Bakke, "Art for Plants' Sake? Questioning Human Imperialism in the Age of Biotech," *Parallax* 18, no. 4 (2012): 10.
27 Quentin Meillassoux, *After Finitude: An Essay on the Necessity of Contingency* (New York: Bloomsbury, 2010).
28 Bachelard, *The Poetics of Space*, 105.
29 Michel Serres, *The Natural Contract*, trans. Elizabeth MacArthur and William Paulson (Ann Arbor: University of Michigan Press, 1995).
30 Karen Barad, *Meeting the Universe Halfway: Quantum Physics and the Entanglement of Matter and Meaning* (Durham: Duke University Press, 2007), 210.
31 Braidotti, *The Posthuman*, 108.
32 Peter Sloterdijk, *Bubbles: Spheres 1* (Los Angeles: Semiotext(e), 2011), 280.
33 Timothy Morton, *Hyperobjects: Philosophy and Ecology After the End of the World* (Minneapolis: University of Minnesota Press, 2013).
34 Monica Bakke, "Art and Metabolic Force in Deep Time Environments," *Environmental Philosophy* 14, no. 1 (2017): 2
35 Deborah Bird Rose, *Wild Dog Dreaming: Love and Extinction* (Charlottesville: University of Virginia Press, 2011), 146.
36 Laurence, interview by Prudence Gibson.
37 Luce Irigaray, *Marine Lover of Friedrich Nietzsche*, trans. Gillian C. Gill (New York: Columbia University Press, 1991), 13.

Works cited

Adamowsky, Natascha. *The Mysterious Science of the Sea, 1775–1943*. London: Routledge, 2016.

Adamson, Glenn. "The Real in the Rococo." *Rethinking the Baroque*, ed. Helen Hills. Surrey and Burlington: Ashgate, 2011. 143–57.

Addison, Joseph, and Richard Steele. *The Spectator*, ed. G. Gregory Smith. 8 vols. bound as 4. London: Dent, 1907.

Adorno, Theodor, Max Horkheimer, and Robert Hullot-Kentor. "Odysseus or Myth and Enlightenment." *New German Critique* 56 (1992): 109–41. doi:10.2307/488331.

Alberti, Leon Battista. *On the Art of Building in Ten Books*, trans. J. Rykwert, N. Leach, and R. Tavernor. Cambridge, MA: MIT Press, 1988.

Albrecht, Glenn et al. "Solastalgia: the distress caused by environmental change." *Australasian Psychiatry* 15 (2007): S95–S98.

Amico, Leonard N. *Bernard Palissy: In Search of Earthly Paradise*. Paris: Flammarion, 1996.

Amos, Jonathan. "Museum's 'Dippy' dinosaur makes way for blue whale." *BBC News*, 29 January 2015. https://www.bbc.com/news/science-environment-31025229.

Andersen, H. C. "The Mermaid," trans. Lucie, Lady Duff Gordon. *Bentley's Miscellany*. Vol. XIX. London: Richard Bentley, 1846. 377-90.

Anderson, Kevin and Alice Bows. "Beyond 'Dangerous' Climate Change: Emission Scenarios for a New World." *Philosophical Transactions of the Royal Society of London A: Mathematical, Physical and Engineering Sciences* 369 (2010): 20–44.

——. "Reframing the climate change challenge in light of post-2000 emission trends." *Philosophical Transactions of the Royal Society of London A: Mathematical, Physical and Engineering Sciences* 366 (2008): 3863–82.

Anderson, Wilda. " Vol au-dessus d'un paradigme nouveau." *Ballons et regards d'en haut*, ed. Michel Delon and Jean Goulemot. Cahiers de littérature française. Vol. 5. Paris: L'Harmattan, 2007. 27-43.

The Argus, 23 August 1896.

Arkush, Michael. "Q&A With Dick Clark: 'Popular Music Is the Soundtrack of Our. . .Lives.' *Los Angeles Times*, 7 February 1994. http://articles.latimes.com/1994-02-07/entertainment/ca-20043_1_dick-clark.

Arnold, Matthew. *Complete Poems*, ed. Kenneth Allott. New York: Longman, 1979.

Ashworth, William. *Theories of the Earth, 1644–1830: The History of a Genre*. Kansas City, MO: Linda Hall Library, 1984.

Auden. W. H. *Collected Poems*. New York: Vintage Reprint, 1991.

——. *The Enchafèd Flood: or, The Romantic Iconography of the Sea*. New York: Vintage Books, 1967.

Austen, Jane. *Jane Austen's Letters*, edn. Deirdre Le Faye. 4th ed. Oxford: Oxford University Press, 2011.

——. *Sanditon*, ed. Margaret Drabble. London: Penguin, 1974.

Auster, Peter J. et al. "Developing an Ocean Ethic: Science, Utility, Aesthetics, Self-Interest, and Different Ways of Knowing." *Conservation Biology* 23.1 (2009): 233–5.

Baarsen, Reinier J. *Paris 1650-1900: Decorative Arts in the Rijksmuseum*. New Haven: Yale University Press, 2013.

Bach, J.P.S. *The Pearling Industry of Australia: An Account of Its Social and Economic Development*. Canberra: Department of Commerce and Agriculture, 1955.

Bachelard, Gaston. *L'Eau et les rêves: essai sur l'imagination de la matière*. Paris: J. Corti, 1942.

——. *The Poetics of Space*. Boston: Beacon, 1994.

Baker, Samuel. *Written on the Water: British Romanticism and the Maritime Empire of Culture*. Charlottesville: University of Virginia Press, 2010.

Bakke, Monika. "Art and Metabolic Force in Deep Time Environments." *Environmental Philosophy* 14.1 (2017): 41–59.

——. "Art for Plants' Sake? Questioning Human Imperialism in the Age of Biotech." *Parallax* 18.4 (2012): 9–25.

Barad, Karen. *Meeting the Universe Halfway: Quantum Physics and the Entanglement of Matter and Meaning*. Durham, NC: Duke University Press, 2007.

Barkan, Leonard. *The Gods Made Flesh: Metamorphosis and the Pursuit of Paganism*. New Haven: Yale University Press, 1986.

Baudelaire, Charles. *Les Fleurs du mal*. Paris: Poulet-Malassis et De Broise, 1857.

Bauman, Zygmunt. *Liquid Modernity*. Cambridge, UK: Polity, 2000.

Bazin, André. "Information or Necrography" [1957]. *André Bazin's New Media*, ed. and trans. Dudley Andrews. Oakland: University of California Press, 2014. 124-5.

The Beatles. "Strawberry Fields Forever." Song. London: Parlophone, 1967.

—. "The Yellow Submarine." Song. London: Parlophone, 1966.

"The Beatles arrive in the Philippines." *BeatlesBible.com*. Accessed 1 May 2016. http://www.beatlesbible.com/1966/07/03/beatles-arrive-in-manila-philippines.

"The Beatles leave Manila, Philippines." *BeatlesBible.com*. Accessed 1 May 2016. http://www.beatlesbible.com/1966/07/05/the-beatles-leave-the-philippines.

Beirut. *The Rip Tide*. Studio album. New York: Pompeii Records, 2011.

Bennett, Jane. *Vibrant Matter: A Political Ecology of Things*. Durham, NC: Duke University Press, 2010.

Béhar, Henri. *Le Surréalisme dans la presse de gauche, 1924–1939*. Paris: Éditions Mélusine, 2012.

Berger, Jr., Harry. *Second World and Green World*. Berkeley: University of California Press, 1988.

Bernardin de Saint-Pierre. *Études de la nature*. Vol. 3. Bâle, 1795.

Blumenberg, Hans. *Shipwreck With Spectator: Paradigm for a Metaphor of Existence*, trans. Steven Rendall. Cambridge, MA: The MIT Press, 1997 [1979].

Berkeley, George. *Essay towards a New Theory of Vision*. London: Dent, 1972.

Blight, Louise, Mark Drever, and Peter Arcese. "A Century of Change in Glaucous-winged Gull (Larus glaucescens) Populations in a Dynamic Coastal Environment." *The Condor* 117.1 (2015): 108–20.

Borrelle, S. B. et al. "Opinion: why we need an international agreement on marine plastic pollution." *Proc. Natl Acad. Sci. USA* 114 (2017): 9994–7.

Bourdier, Franck. "Lamarck et Geoffroy Saint-Hilaire face au problème de l'évolution biologique." *Revue d'histoire des sciences* 25.4 (1972): 311–325. Persée. doi: 10.3406/rhs.1972.3305.

Bowen, Brian W. et al. "The Origins of Tropical Marine Biodiversity." *Trends in Ecological Evolution* 28 (2013): 359–66.

Boyle, Robert. "The Temperature of the Subterranean and submarine Regions, As to Heat and Cold." *The Philosophical Works*, ed. Peter Shaw. Vol. 3. London: Innys et al., 1725. 232–49.

Braidotti, Rosi. *The Posthuman*. Cambridge, UK: Polity, 2013.

Brandt, Bill. "Au cimitière des anciennes galères. Scilly Isles, Angleterre." *Minotaure* 6 (Winter 1935): 4.

Bredekamp, Horst. *The Lure of Antiquity and the Cult of the Machine: The Kunstkammer and the Evolution of Nature, Art and Technology*, trans. Allison Brown. Princeton: Markus Wiener Publishers, 1995.

Brenner, Neil and Christian Schmid. "Planetary Urbanization." *Urban Constellations*, ed. Matthew Gandy. Berlin: Jovis, 2012. 10–13.

Breton, André. *Break of Day*, trans. Mark Pollizzotti and Mary Ann Caws. Lincoln, NE: University of Nebraska Press, 1999.

——. *Lettres à Simone Kahn*. Paris: Gallimard, 2016.

——. *Oeuvres complètes*, eds. Marguerite Bonnet and Etienne-Alain Hubert. Vol. 1. Paris: Gallimard, Bibliothèque de la Pléiade, 1988.

——. *Oeuvres complètes*, eds. Marguerite Bonnet and Etienne-Alain Hubert. Vol. 4. Paris: Gallimard, Bibliothèque de la Pléiade, 2008.

——. *Soluble Fish. Manifestos of Surrealism*, trans. Richard Seaver and Helen R. Lane. Ann Arbor: University of Michigan Press, 1972. 63–4.

——. *Surrealism and Painting*, trans. Mark Pollizzotti. Boston: MFA Publishing, 2002.

Breton, André, André Parinaud et al. *Conversations: The Autobiography of Surrealism*, trans. Mark Polizzotti. New York: Marlowe & Company, 1993.

The Brisbane Courier, 31 October 1893.

Brunon, Hervé and Monique Mosser. *L'Imaginaire des grottes dans les jardins européens*. Paris: Hazan, 2014.

Bullen, J. Barrie. "Alfred Waterhouse's Romanesque 'Temple of Nature': The Natural History Museum, London." *Architectural History* 49 (2006): 257–8.

Buffon, Georges-Louis Leclerc de. *Histoire naturelle, générale et particulière, avec la description du Cabinet du Roy*. Vol. 1. Paris: Impriméric royale, 1749.

Burke, Edmund. *A Philosophical Enquiry into the Origin of our Ideas of the Sublime and Beautiful*, ed. Adam Phillips. Oxford: Oxford University Press, 1998.

Byron, George Gordon, Lord. *Byron*, ed. Jerome J. McGann. Oxford: Oxford University Press, 1986.

——. *Byron's Letters and Journals*, ed. Leslie A. Marchand. 13 vols. London: John Murray, 1973–94.

——. "Swimming across the Hellespont: Letter from the Right Honourable Lord Byron to Mr Murray," *The London Magazine* 3 (April 1821): 363–5.

Callicott, J. Baird. "La Nature est morte, vive la nature!" *The Hastings Center Report* 22.5 (1992): 16–23.

Cameron, James. "Ghost Walking in Titanic." *National Geographic* 221.4 (April 2012): 100–1. http://www.nationalgeographic.com/magazine/2012/04/titanic-shipwreck-underwater-exploration-james-cameron/.

Capstick, Stuart B., Nick F. Pigeon, Adam J. Corner, Elspeth M. Spence and Paul N. Pearson. "Public understanding in Great Britain of ocean acidification." *Nature Climate Change* 6 (2016): 763–7.

Carson, Rachel L. *The Sea Around Us*. London: Penguin Books, 1956 [1951].

Catts, Oron, ed. *The Aesthetics of Care?: The artistic, social and scientific implications of the use of biological/medical technologies for artistic purposes*. Nedlands, W.A.: SymbioticA, 2002.

Caus, Salomon de. *Les Raisons des forces mouvants avec divers Machines tant utilles que plaisants aus quelles sont adioinct plusieurs desseings de grottoes et fontaines*. Frankfurt: Jan Norton, 1615.

Cazzato, Vincenzo, Marcello Fagiolo, and Maria Adriana Giusti, eds. *Atlante delle grotte e dei ninfei in Italia*. Milan: Electra, 2001.

Cenci, Elena. "'L'antro delle ninfe': significato ed evoluzione dall'Antichità al Cinquecento." *Il Giardino Storico All'Italiana*, ed. Francesco Nuvolari. Milan: Electra, 1992. 147–53.

Chappell, David A. *Double Ghosts: Oceanian Voyagers on Euroamerican ships*. New York and London: M.E. Sharpe, 1997.

Cheever, John. *Collected Stories and Other Writings*. New York: Library of America, 2009.

"Circumstances of the Grounding: 'An uncanny sensation.'" *Northern Herald* [Queensland], 14 July 1926, 38.

Clarke, David. *Water and Art*. London: Reaktion Books, 2010.

Clayson, Allan. "'Wish You were Here': The Significance of Coleridge's Holidays at Ramsgate." *The Coleridge Bulletin*, n.s., 16 (2000): 15–23.

Cleary, Scott. "The Ethos Aquatic: Benjamin Franklin and the Art of Swimming." *Early American Literature* 46.1 (2011): 51–67.

Cocker, Emma. "Desiring to be Led Astray." *Papers of Surrealism* 6 (2007): 1–30.

Coffin, David R. *The Villa in the Life of Renaissance Rome*. Princeton: Princeton University Press, 1979.

Cohen, Margaret. "The Chronotopes of the Sea." *The Novel*, ed. Franco Moretti. Vol. 2. Princeton: Princeton University Press, 2006. 647–67.

——. "Denotation in Alien Environments: The Underwater *Je Ne Sais Quoi*." *Representations* 125.1 (2014): 103–26.

——. *The Novel and the Sea*. Princeton: Princeton University Press, 2010.

——. "Underwater Optics as Symbolic Form." *French Politics, Culture & Society* 32 (2014): 1–23.

Cohn, Norman. *Noah's Flood: The Genesis Story in Western Thought*. New Haven, CT: Yale University Press, 1996.

Coleridge, Samuel Taylor. *The Collected Letters of Samuel Taylor Coleridge*, ed. E. L. Griggs. 6 vols. Oxford: Oxford University Press, 1956–71.

——. *The Collected Works of Samuel Taylor Coleridge*. 3 vols. in 6 parts. Princeton: Princeton University Press, 2001.

——. *Notebooks*, ed. Kathleen Coburn. 4 vols. London: Routledge, 2002.

Condillac, Étienne de. *Logic*, trans. W. R. Albury. New York: Abaris Books, 1980.

Conti, Natale. *Mythologiae*, trans. John Mulryan and Steven Brown. Vol. 1. Tempe, AZ: Arizona Center for Medieval and Renaissance Studies, 2006.

Cook, James. *The Journals of Captain James Cook: The Voyage of the Endeavour, 1768–1771*, ed. J. C. Beaglehole. Cambridge, UK: Cambridge University Press for the Hakluyt Society, 1968.

Corbin, Alain. *The Lure of the Sea: The Discovery of the Seaside in the Western World, 1750–1840*, trans. Jocelyn Phelps. Berkeley: University of California Press, 1994.

Country Life Illustrated 3 (19 March 1898).

Cousteau, Jacques-Yves. *Epaves*. Film. France: Jacques-Yves Cousteau producer, 1946 [1943].

——. with Frédéric Dumas. The Silent World. New York: Harper & Row, 1953.

——. and Louis Malle. *Le Monde du silence*. Film. France: FSJYC Production, 1956.

Cowper, William. *The Poems*. New York: Charles Wells, 1835.

Crawford, Jason. *Allegory and Enchantment: An Early Modern Poetics*. Oxford: Oxford University Press, 2017.

Crawford, Robert. *Young Eliot*. New York: Farrar, Straus and Giroux, 2015.

Crutzen, Paul and Will Steffen. "How Long Have We Been In The Anthropocene Era? An Editorial Comment." *Climatic Change* 61 (2003): 253.

Culler, A. Dwight. *Imaginative Reason: The Poetry of Matthew Arnold.* New Haven: Yale University Press, 1966.

Daniels, Stephen. "Suburban pastoral: Strawberry Fields forever and Sixties memory." *Cultural Geographies* 13.1 (2006): 28–54.

Darwin, Charles. *The Autobiography of Charles Darwin, 1809–82*, ed. Nora Barlow. New York and London: Norton, 1958.

——. *On the Origin of Species by Means of Natural Selection, Or, The Preservation of Favoured Races in the Struggle for Life.* London: J. Murray, 1859.

——. *The Structure and Distribution of Coral Reefs.* London: Ward, Lock & Co., 1910.

Darwin, Erasmus. *Zoonomia.* 3 vols. London: J. Johnson, 1801.

Davy Jones's Locker, Entry, The National Maritime Museum. Accessed May 2018. http://collections.rmg.co.uk/collections/objects/561491.html

Deakin, Roger. *Waterlog: A Swimmer's Journey through Britain.* London: Chatto & Windus, 1999.

Descartes, René. *Treatise on Light. The World and other Writings*, ed. and trans. Stephen Gaukroger. Cambridge, UK: Cambridge University Press, 1998.

Desmarest, Nicolas. *Conjectures physico-mechaniques sur la propagation des secousses dans les tremblemens de terre: et sur la disposition des lieux qui en ont ressenti les effets.* Paris, 1756.

Desmond, Adrian. *Huxley: From Devil's Disciple to Evolution's High Priest.* Reading: Addison-Wesley, 1997.

Desnos, Robert. "Confession d'un enfant du siècle." *La Révolution Surréaliste* 6 (March 1926): 21–2.

Di Cioccio, Davide. "Ecology of the toxic dinoflagellate Ostreopsis cf. ovata along the coasts of the Campania region (Tyrrhenian Sea, Mediterranean Sea)." PhD Thesis, Università degli Studi di Napoli Federico II, 2014.

Dickens, Charles. *The Mystery of Edwin Drood.* London: Folio Society, 1982.

Diderot, Denis, and Jean le Rond D'Alembert, eds. *Encyclopédie, ou dictionnaire raisonné des sciences, des arts et des métiers.* ARTFL *Encyclopédie Project*, ed. Robert Morrissey. Chicago: University of Chicago, n.d. Accessed May 2018. http://encyclopedie.uchicago.edu/.

Doney Scott C. et al. "Ocean Acidification: The Other CO2 Problem." *Annual Review of Marine Science* 1 (2009), 169–92.

Doran, Robert. *The Theory of the Sublime from Longinus to Kant.* Cambridge, UK: Cambridge University Press, 2015.

Dryden, John. *The Works of John Dryden, Vol. 3*: Poems 1685-1692. Ed. Earl Roy Miner and Vinton A. Dearing. Berkeley: University of California Press, 1969.

Elderkin, G.W. "The Natural and the Artificial Grotto." *The Journal of the American School of Classical Studies at Athens* 10.2 (1941): 125–37.

Eldredge, Charles C. "Wet Paint: Herman Melville, Elihu Bedder and Artists Undersea." *American Art* 11.2 (1997): 107–35.

Elias, Ann. "Sea of Dreams: André Breton and the Great Barrier Reef." *Papers of Surrealism* 10 (Summer 2013): 1–15.

Eliot, T. S. *Collected Poems, 1909-1962.* New York: Harcourt, Brace & World, 1970.

——. *The Poems of T. S. Eliot: Volume I, Collected and Uncollected Poems*, eds. Christopher Ricks and Jim McCue. Baltimore: Johns Hopkins University Press, 2015.

——. *Selected Prose of T. S. Eliot*, ed. Frank Kermode. London: Faber and Faber, 1975.

——. *The Use of Poetry and the Use of Criticism.* London: Faber and Faber, 1933.

——. *The Varieties of Metaphysical Poetry*, ed. R. S. Faber. London: Faber and Faber, 1993.

Elliott, Anthony. *The Mourning of John Lennon*. Berkeley: University of California Press, 1999.

Ellis, Richard. "The Lower Depths: *Song For The Blue Ocean: Encounters Along the World's Coasts and Beneath the Seas* by Carl Safina." *Los Angeles Times Book Review*, 18 January 1998. http://articles.latimes.com/1998/jan/18/books/bk-9463.

Endt-Jones, Marion. "'Something Rich and Strange': Coral in Contemporary Art." *Framing the Ocean, 1700 to the Present: Envisaging the Sea as Social Space*, ed. Tricia Cusack. Farnham and Burlington: Ashgate, 2014. 223–38.

Eucalyptus. "Farm & Garden." *Townsville Daily Bulletin* [Queensland], 26 January 1932, 7.

Exton, John. *The maritime dicæologie; or, sea-jurisdiction of England*. London: 1746.

Findlen, Paula. "The Museum: Its Classical Etymology and Renaissance Genealogy." *Journal of the History of Collections* 1.1 (1989): 59–78.

Firebrace, William. "Eyes Aquatic." *AA Files* 62 (2011): 47–62.

Foucault, Michel. *The Order of Things: An Archaeology of the Human Sciences*. New York: Routledge, 1970.

Fouquet, Roger and Peter Pearson. "Seven Centuries of Energy Services: The Price and Use of Light in the United Kingdom (1300-2000)." *The Energy Journal* 27.1 (2006): 139–77.

Franklin, Benjamin. *Experiments and Observations on Electricity, Made at Philadelphia in America*. London, 1769.

Fraser Jenkins, A.D. "Cosimo de' Medici's Patronage of Architecture and the Theory of Magnificence." *Journal of the Warburg and Courtauld Institutes* 33 (1970): 162–70.

Fretz, Lauren. "Surréalisme sous-l'eau: Science and Surrealism in the Early Films and Writings of Jean Painlevé." *Film and History* 40.2 (Fall 2010): 45–65.

Freud, Sigmund. "The Uncanny" (1919). *The Standard Edition of the Complete Psychological Works of Sigmund Freud*, ed. and trans. John Strachey. Vol. 17. London: Hogarth Press, 1964. 217–56.

Friedman, Norman. "The Young Matthew Arnold 1847–1849: 'The Strayed Reveller' and 'The Forsaken Merman'" *Victorian Poetry* 9.4 (Winter 1971): 405–28.

Gallagher, Ryan. "From Radio to Porn, British Spies Track Web Users' Online Identities." *The Intercept*, 25 September 2015. https://theintercept.com/2015/09/25/gchq-radio-porn-spies-track-web-users-online-identities

Ganter, Regina. *The Pearl Shellers of Torres Strait: Resource Use, Development and Decline, 1860s–1960s*. Melbourne: Melbourne University Press, 1994.

Gattuso, Jean-Pierre and Lina Hansson, L. *Ocean Acidification*. Oxford: Oxford University Press, 2011.

Gelcich, S., Buckley, P., Pinnegar, J.K. et al. "Public Awareness, Concerns, and Priorities about Anthropogenic Impacts on Marine Environments." *Proceedings of the National Academy of Sciences of the United States of America*, 111.42 (2014): 15042-7

Gibson, James J. *The Ecological Approach to Visual Perception*. Boston: Houghton Mifflin, 1979.

Gibson, Prudence. "The Colour Green." *Covert Plants: Vegetal Consciousness and Agency in an Anthropocentric World*, eds. Gibson and Baylee Brits. San Diego: Punctum, 2018. n.p.

—. *Janet Laurence: The Pharmacy of Plants*. Sydney: New South Publishing, 2015.

—. *The Rapture of Death*. Sydney: Boccalatte Books, 2010.

Gibson, Prudence and Monica Gagliano "The Feminist Plant: Changing Relations with the Water Lily." *Ethics and the Environment* 22.2 (2017): 125–46.

Gilpin, William. *Observations on the River Wye, and Several Parts of South Wales, &c.* London: R. Blamire, 1782.

Gleeson White, Jane. "It's only natural: the push to give rivers, mountains and forests legal rights." *The Guardian*, 1 April 2018. https://www.theguardian.com/australia-news/2018/apr/01/its-only-natural-the-push-to-give-rivers-mountains-and-forests-legal-rights.

Goethe, Johann Wolfgang von. *Goethe's Theory of Colours*, trans. Charles Lock Eastlake. London: John Murray, 1810; reprinted Cambridge, UK: Cambridge University Press, 2014.

Gossman, Lionel. "Michelet and Natural History: The Alibi of Nature." *Proceedings of the American Philosophical Society* 145.3 (2001): 283–333.

Gould, Stephen Jay. *Ontogeny and Phylogeny*. Cambridge, MA: The Belknap Press, 1977.

Greif, Mark. "Radiohead, Or the Philosophy of Pop." *Radiohead and Philosophy: Fitter, Happier, More Deductive*, eds. Brandon Forbes and George Reisch. Chicago: Open Court, 2009.

Hamera, Judith. *Parlor Ponds: The Cultural Work of the American Home Aquarium, 1850–1970*. Ann Arbor: University of Michigan Press, 2011.

Hamilton, Clive. "No Escaping the Blue Marble." *The Conversation*, 15 August 2015. https://theconversation.com/no-escaping-the-blue-marble-46454.

Haq, Gary and Alistair Paul. *Environmentalism Since 1945*. London: Routledge, 2013.

Hansen, James. "Climate Urgency." *Simulation & Gaming* 44 (2013): 232–43.

Haraway, Donna. "Anthropocene, Capitalocene, Plantationocene, Chthulucene: Making Kin." *Environmental Humanities* 6 (2015): 159–65.

Harrison, Anthony J. *Savant of the Australian Seas, William Saville-Kent, 1842–1908*. 2nd edn. Hobart: Tasmanian Historical Research Association, 1997. http://www.stors.tas.gov.au/au-7-0074-00017.

Harrison, Peter. *The Bible, Protestantism, and the Rise of Natural Science*. Cambridge, UK: Cambridge University Press, 1998.

Hay, Pete. "What the Sea Portends: A Reconsideration of Contested Island Tropes." *Island Studies Journal* 8.2 (2013): 209–32.

Helmreich, Stefan. *Alien Ocean: Anthropological Voyages in Microbial Seas*. Berkeley: University of California Press, 2009.

Hirst, Damien et al. *Treasures from the Wreck of the Unbelievable*. London: Other Criteria, 2017.

Hirst, Damien, Christoph Gerigk and Steve Russell. *The Undersea Salvage Operation: Treasures from the Wreck of the Unbelievable*. London: Other Criteria, 2017.

Hirst, Damien and Elena Geuna, curators. *Treasures from the Wreck of the Unbelievable*. Exhibition. Palazzo Grassi and Punta della Dogana, 9 April–3 December 2017.

Hoare, Philip. *The Sea Inside*. London: Fourth Estate, 2014.

Hobbes, Thomas. *Elements of Philosophy*. *The Works of Thomas Hobbes*, ed. Sir William Molesworth. 16 vols. London: John Bohn, 1839.

Hogarth, William. *The Analysis of Beauty*. London: J. Reeves, 1753.

Holcroft, Thomas. *The Adventures of Hugh Trevor*, ed. Seamus Deane. Oxford: Oxford University Press, 1973.

Hönisch, Bärbel et al. "The Geological Record of Ocean Acidification." *Science* 335.6072 (2012): 1058–63.

Horace. *The Odes of Horace*, trans. David Ferry. New York: Farrar, Straus and Giroux, 1997.

Horkheimer, Max and Theodor W. Adorno. *Dialectic of Enlightenment: Philosophical Fragments*, ed. Gunzelin Schmid Noerr, trans. Edmund Jephcott. Stanford: Stanford University Press, 2002.

Hooke, Robert. *Posthumous Works*, ed. Richard Waller. London: Samuel Smith and Benjamin Walford, 1705.

Horsbrugh, Becky. "The Hellespont Swim: Following in Byron's Wake." *The Guardian*, 6 May 2010.

Hugo, Victor. *Oeuvres Complètes de Victor Hugo*. Paris: Eugene Renduel, 1835.

Hume, David. *A Treatise of Human Nature*, eds. L.A. Selby-Bigge and P.H. Nidditch. Oxford: Clarendon Press, 2009.

Hurley, Frank. "Being an account of a voyage to the Barrier Reef. The Torres Straits Islands and New Guinea. From 2 December 1920. Islands visited, Darnley, Murray, Mabuiag, Moa." Papers of Frank Hurley, 1912–1962. National Library of Australia, MS 883 Series 1, item 7.

——. "The Floor of the Coral Sea: Among the Blue-green Opal." *Sun* [New South Wales], 7 February 1926, 17.

——. "Frank Hurley's diary, when on tour with the Shackleton Picture, 21 November 1919– 20 January 1920." Papers of Frank Hurley 1912–1962. National Library of Australia, MS 888, Series 1, item 6.

——. "A Great Enterprise: Filming the Great Barrier Reef." *The Richmond River Herald and Northern Districts Advertiser* [New South Wales], 7 December 1920, 1.

——. "The Imperial Trans-Antarctic Expedition Diary." *The Diaries of Frank Hurley, 1912–1941*, eds. Robert Dixon and Christopher Lee. London and New York: Anthem Press, 2011. 11-58.

——. *Pearls and Savages: Adventures in the Air, on Land and Sea – in New Guinea*. New York and London: Putnam, 1925.

——. "Where Icy Blizzards Blow." *Sun*, 26 June 1927, 22.

Ibarra, Yaadira et al. "Microfacies of the Cotham Marble: A tubestone carbonate microbialite from the Upper Triassic, South Western UK." *Palaios* 29.1 (2013): 1–15.

Ingold, Tim. *The Perception of the Environment: Essays on livelihood, dwelling and skill*. London and New York: Routledge, 2000.

Innis, Michelle. "Great Barrier Reef Facing Multiple Threats, Report Says." *The New York Times*, 2 December 2016, A6.

—. "Inspecting Reefs From 28,000 Feet Above." *The New York Times*, 27 September 2016, A4.

Irigaray, Luce. *Marine Lover of Friedrich Nietzsche*, trans. Gillian C. Gill. New York: Columbia University Press, 1991.

Irigaray, Luce and Michael Marder. *Through Vegetal Being*. New York: Columbia University Press, 2016.

Isensee, Kirsten and Luis, Valdes *GSDR 2015 Brief: Marine Litter: Microplastics*. IOC/ UNECSCO, 2015.

Ivanina, Anna V. and Inna M Sokolova. "Interactive Effects of Metal Pollution and Ocean Acidification on Physiology of Marine Organisms." *Current Zoology* 61:4 (1 August 2015): 653–668. https://doi.org/10.1093/czoolo/61.4.653.

James, William. *Principles of Psychology*. 2 vols. New York: Henry Holt, 1890.

——. *The Varieties of Religious Experience: A Study in Human Nature*. New York: Routledge, 2008.

Januszczak, Waldemar, dir. *Rococo: Travel, Pleasure, Madness*. Film. London: British Broadcasting Corporation, 2014.

Jay, Martin. "Scopic Regimes of Modernity." *Vision and Visuality*, ed. Hal Foster. Seattle: Bay Press, 1988. 3–23.

Johnston, David, and Mark Mazetti. "Interrogation Inc.: A Window Into C.I.A.'s Embrace of Secret Jails." *The New York Times*, 12 August 2009. https://www.nytimes.com/2009/08/13/world/13foggo.html.

Jukić, Nina. "Pop, Sampling and Postmodernism: The Case of Radiohead's Idioteque." *Proceedings of the 2nd International Conference of Students of Systematic Musicology*, Ghent, Belgium, November 18–20, 2009. 75–80.

Kant, Immanuel. *The Critique of Judgment*, trans. James Creed Meredith. Oxford: Clarendon Press, 1952.

Kaplan, Edward. "Michelet évolutionniste." *Romantisme* 10 (1975): 111–28. Persée. doi: 10.3406/roman.1975.5003.

Kawaguchi, So et al. "Risk Maps for Antarctic Krill under Projected Southern Ocean Acidification." *Nature Climate Change* 3 (2013): 843–47.

Keats, John. *Letters of John Keats*, ed. Robert Gittings. Oxford: Oxford University Press, 1970.

——. *The Poems of John Keats*, ed. Jack Stillinger. London: Heinemann, 1978.

Kemp, Martin. "Wrought by No Artist's Hand: The Natural, the Artificial, the Exotic, and the Scientific in Some Artefacts from the Renaissance." *Reframing the Renaissance: Visual Culture in Europe and Latin* America *1450–1650*, ed. Claire Farago. New Haven: Yale University Press, 1995. 177–98.

Kolbert, Elizabeth. *The Sixth Extinction*. New York: Henry Holt and Company, 2014.

Kris, Ernst. *Le Style rustique*, trans. Christophe Jouanlanne. Paris: Macula, 2005.

Kyle, Noeline. *A Greater Guilt: Constance Emilie Kent and the Road Murder*. Brisbane: Boolarong, 2009.

Lambert, David, Luciana Martins and Miles Ogborn. "Currents, Visions and Voyages: Historical Geographies of the Sea." *The Journal of Historical Geography* 32 (2006): 479–93.

Lamberton, Robert. *On The Cave of the Nymphs*. Barrytown, NY: Station Hill Press, 1983.

Larson, Jennifer. *Greek Nymphs: Myth, Cult, Lore*. Oxford: Oxford University Press, 2001.

Laurence, Janet. Interview by Prudence Gibson. 28 December 2017.

Le Dû-Blayo, Laurence and Olivier Musard. "Towards a Shared Language: Semantic Exchanges and Cross-disciplinary Interaction." Introduction to *Underwater Seascapes: From geographical to ecological perspectives*, eds. Musard, Laurence Le Dû-Blayo et al. Cham: Springer, 2014. 1–15.

Lingis, Alphonso. "The Rapture of the Deep." *Excesses: Eros and Culture*. Albany, NY: State University Press of New York, 1983. 1-16.

——. *Sensations: Intelligibility in Sensibility*. New York: Humanity Books, 1996.

Lippit, Akira M. *Electric Animal: Toward a Rhetoric of Wildlife*. Minneapolis: University of Minnesota Press, 2000.

Locke, John. *An Essay Concerning Human Understanding*, ed. Peter H. Nidditch. Oxford: Clarendon Press, 1979.

Love, Rosaleen. *Reefscape. Reflections on the Great Barrier Reef.* Washington: Joseph Henry Press, 2001.

Luc, Jean-André de. *Lettres physiques et morales, sur les montagnes et sur l'histoire de la terre et de l'homme*. En Suisse: Chez les Libraires Associés, 1778.

Luria, S. M. and Jo Ann S. Kinney. "Underwater Vision." *Science* 167.3924 (1970): 1454–61.

Lyell, Charles. *Principles of Geology: Being an Attempt To Explain the Former Changes of the Earth's Surface, By Reference to Causes Now in Operation*. London: John Murray, 1830–35.

Mabille, Pierre. "Preface in Praise of Popular Prejudices," trans. Michael Richardson and Krzysztof Fijalkowski. *The Surrealism Reader: An Anthology of Ideas*, eds. Dawn Ades and Michael Richardson. London: Tate Publishing, 2015. 32–40.

MacCarthy, Fiona. *Byron: Life and Legend*. London: John Murray, 2002.

Magruder, Kerry V. "Thomas Burnet, Biblical Idiom, and Seventeenth-Century Theories of the Earth." *Nature and Scripture in the Abrahamic Religions: Up to 1700*, eds. Jitse M. van der Meer and Scott Mandelbrote. Vol. 2. Leiden: Brill, 2008. 451–90.

Maillet, Benoît de. *Telliamed: ou, entretiens d'un philosophe indien avec un missionaire françois sur la diminution de la mer, la formation de la terre, l'origine de l'homme, &c.* Amsterdam: Chez l'Honoré & Fils, 1748.

Malle, Louis. *Malle on Malle*, ed. Philip French. New York: Faber and Faber, 1993.

"Man's Fight with Nature: Professor Ray Lankester's New Book, 'The Kingdom of Man.'" *The Daily News* [Perth], 23 March 1907, 13.

Margulis, Lynn. *Symbiotic Planet: A New Look at Evolution.* New York: Basic Books, 1998.

Marsigli, Luigi Ferdinando. *Histoire physique de la mer. Ouvrage enrichi de figures dessinés d'après le naturel.* Amsterdam, 1725.

McCalman, Iain. *Darwin's Armada.* Melbourne: Penguin, 2009.

——. *The Reef: A Passionate History.* Melbourne: Viking, 2013; New York: Scientific American/Farrar, Straus and Giroux, 2014.

McNab, Robert. *Ghost Ships: A Surrealist Love Triangle.* New Haven and London: Yale University Press, 2004.

Medwin, Thomas. *Conversations of Lord Byron: Noted during a Residence with his Lordship at Pisa, in the Years 1821 and 1822.* London: H. Colburn, 1824.

Meillassoux, Quentin. *After Finitude: An Essay on the Necessity of Contingency.* New York: Bloomsbury, 2010.

Mentz, Steve, and Martha Elena Rojas, eds. *The Sea and Nineteenth-Century Anglophone Literary Culture.* New York: Routledge, 2017.

Mentz, Steve. *At the Bottom of Shakespeare's Ocean.* New York: Continuum, 2009.

——. *Shipwreck Modernity: Ecologies of Globalization 1550–1719.* Minneapolis: University of Minnesota Press, 2015.

Mercado, Gustave. *The Filmmaker's Eye.* Burlington, MA: The Focal Press, 2011.

Merleau-Ponty, Maurice. *Nature: Course Notes from the Collège de France*, trans. Robert Vallier. Evanston, IL: Northwestern University Press, 2003.

Michelet, Jules. *La Mer.* 2nd edn. Paris: Hachette, 1861.

—-. *The Sea.* New York: Rudd and Carleton, 1861.

Middleton, Christopher. *A Short introduction for to Learne to Swimme.* London, 1595.

Milam, Jennifer. "Play Between Disciplines: The Problem of the Ludic in Rococo Art and Enlightenment Culture." *The Interdisciplinary century: tensions and convergences in eighteenth-century art, history and literature*, eds. Julia V. Douthwaite and Mary Vidal. Oxford: Voltaire Foundation, 2005. 102–13.

Miller, Naomi. *Heavenly Caves: Reflections on the Garden Grotto.* New York: G. Braziller, 1982.

Morel, Philippe. "Mannerist Grottos in Sixteenth-Century Italy." *Sixteenth-Century Italian Art*, ed. Michael W. Cole. Malden, MA: Blackwell, 2006. 115–34.

Molesworth, Jesse. "Gothic Time, Sacred Time." *Modern Language Quarterly* 75.1 (2014): 29–55.

Montaigne, Michel de. *Journal de voyage*, ed. Fausta Garavini. Paris: Gallimard, 1983.

Morton, Timothy. *Hyperobjects: Philosophy and Ecology After the End of the World.* Minneapolis: University of Minnesota Press, 2013.

Morton-Evans, Christine and Michael Morton-Evans. *The Flower Hunter: The Remarkable Life of Ellis Rowan.* Sydney: Simon & Schuster, 2008.

NASA. "Earth Observatory." Accessed August 2018. https://earthobservatory.nasa.gov/

"Natural History Museum unveils Dippy's replacement, a giant blue whale called Hope." *The Telegraph*, 13 July 2017. https://www.telegraph.co.uk/news/2017/07/13/natural-history-museum-unveils-dippys-replacement-giant-blue/.

Newton, Isaac. *Opticks.* New York: Prometheus Books, 2003.

Nixon, Rob. *Slow Violence and the Environmentalism of the Poor.* Cambridge, MA: Harvard University Press, 2013.

Observations on Indecent Sea-Bathing, as Practised at Different Watering-Places on the Coasts of this Kingdom. London: J. Hatchard, 1805.

Orr, J. C. et al. "Anthropogenic Ocean Acidification Over the Twenty-First Century and Its Impact on Calcifying Organisms." *Nature* 437 (2005): 681–6.

Orr, Linda. *Jules Michelet: Nature, History, and Language*. Ithaca, NY: Cornell University Press, 1976.

Palissy, Bernard. *Oeuvres complètes*, eds. Keith Cameron, Jean Céard, Marie-Madeleine Fragonard, Marie-Dominique Legrand, Frank Lestringant, and Gilbert Schrenk. Paris: Honoré Champion, 2010.

Parr, Susie. *The Story of Swimming: A Social History of Bathing in Britain*. Stockport: Dewi Lewis Media, 2011.

Percey, William. *The Art of Swimming Made Safe and Easy*. London, c.1830.

—. *The Compleat Swimmer: Or, the Art of Swimming*. London, 1658.

Petitier, Paule. *La Géographie de Michelet*. Paris: Editions L'Harmattan, 1997.

Peyssonnel, Jean-André. *Traduction d'un Article des Transactions sur le Corail*. London: 1756.

—. "Traité du Corail ... Pour server a l'histoire naturelle de la mer." Manuscript. British Library. Add MS 4219, ff. 30–1.

"Philippe Tailliez: A la Mémoire du Commandant Philippe Tailliez, Père de la Plongée sous-marine autonome." Accessed May 2018. http://www.philippe.tailliez.net/naissance-du-gers-et-des-premiers.

"Pier Pressure." *The Economist*, 25 November 2017, 38.

Plumwood, Val. *Feminism and the Mastery of Nature*. London: Routledge, 1993.

Poloczanska, E. S. et al. "Responses of marine organisms to climate change across oceans." *Front. Mar. Sci*. 3:62 (2016). doi: 10.3389/fmars.2016.00062.

Pollan, Michael. *The Botany of Desire: A Plant's-Eye View of the World*. New York: Random House, 2002.

Ponge, Francis. "Entretien avec Breton et Reverdy." *Le grand recueil, II. Méthodes*. Paris: Gallimard, 1961.

Prévert, Jacques. "Mort d'un monsieur." *Un Cadavre* (1930): 1.

Quiggin, Alison Hingston. *Haddon, The Head Hunter: A Short Sketch of the Life of AC Haddon*. Cambridge, UK: Cambridge University Press, 1942.

Radcliffe, Ann. *The Mysteries of Udolpho*. New York: Penguin, 2008.

Radiohead. "Bloom." Song. London: XL, 2011.

—. "Idioteque." Song. London: Parlophone, 2000.

—. "No Surprises." Song. London: Parlophone, 1997.

Reid, Thomas. *Directions for Warm and Cold Sea-Bathing; with Observations on their Application and Effects in Different Diseases*. 2nd edn. London: Cadell and Davies, 1798.

Review of *Davy Jones's Locker*. *The Times*, 3 May 1890. In entry, National Maritime Museum. Accessed May 2018. http://collections.rmg.co.uk/collections/objects/561491.html

Rich, Adrienne. "Diving into the Wreck." *Diving Into the Wreck: Poems 1971–1972*. New York, W.W. Norton, 1973. 22–4.

Richardson, Joan. *A Natural History of Pragmatism: The Fact of Feeling from Jonathan Edwards to Gertrude Stein*. New York: Cambridge University Press, 2007.

"R.M.S. Titanic, Guidelines." NOAA Office of General Counsel. Accessed May 2018. https://www.gc.noaa.gov/gcil_titanic-guidelines.html.

Robinson, Tim. *Connemara: The Last Pool of Darkness*. Dublin: Penguin, 2008.

Rodini, Elizabeth. "Baroque Pearls." *Art Institute of Chicago Museum Studies* 25.2 (2000): 68–71.

Roe, Nicholas. *John Keats*. New Haven: Yale University Press, 2012.

Roger, Jacques. *Buffon. Un philosophe au jardin du roi*. Paris: Fayard, 1989.

Rose, Deborah Bird. *Wild Dog Dreaming: Love and Extinction*. Charlottesville: University of Virginia Press, 2011.

Roux, Georges. "Le Val des Muses, et les Muses chez les auteurs anciens." *Bulletin de correspondance hellénique* 78 (1954): 22–48.

Rudwick, Martin. *Bursting the Limits of Time: The Reconstruction of Geohistory in the Age of Revolution*. Chicago: University of Chicago Press, 2005.

Ruskin, John. *Modern Painters*, ed. David Barrie. New York: Knopf, 1987.

Russo, Elena. *Styles of Enlightenment: Taste, Politics, and Authorship in Eighteenth-Century France*. Baltimore: Johns Hopkins University Press, 2007.

Safina, Carl. *Song for the Blue Ocean: Encounters Along the World's Coasts and Beneath the Seas*. New York: Holt Paperbacks, 1999.

Saisselin, Rémy G. *The Enlightenment Against the Baroque: Economics and Aesthetics in the Eighteenth Century*. Berkeley: University of California Press, 1992.

Samuel, Henry. "Row erupts in France over famed sea explorer Jacques Cousteau's 'disgusting' abuse of marine life." *The Telegraph*, 8 July 2015. http://www.telegraph.co.uk/news/worldnews/europe/france/11726747/Row-erupts-in-France-over-famed-sea-explorer-Jacques-Cousteaus-disgusting-abuse-of-marine-life.html.

Saville-Kent, William. *The Great Barrier Reef of Australia*. London: W.H. Allen, 1893; reprinted Provo: Repressed Publishing, 2015.

—. Letter to Sir William Flower, 27 January 1891. Sir William Flower. Semi-Official Papers, Australia and New Zealand, National Library of Australia, MS. M2843.

—. *A Manual of the Infusoria: Including a Description of All Known Flagellate, Ciliate and Tentaculiferous Protozoa*. 3 vols. London: David Bogue, 1880–1881.

—. *The Naturalist in Australia*. London: Chapman Hall, 1897.

—. "Preliminary observations on a natural history collection made on the surveying cruise of HMS Myrmidon." *Papers and Proceedings of the Royal Society of Queensland* 6 (1889): 219–34.

Schama, Simon. *Landscape and Memory*. London: Fontana Press, 1996.

Scott, Walter. *The Poetical Works of Sir Walter Scott*, ed. J. Logie Robertson. New York: Oxford University Press, 1957.

Scruton, Roger. *Beauty: A Very Short Introduction*. Oxford: Oxford University Press, 2011.

"Seagulls Feel the Benefits of Climate Change." *Phys.org*, 5 July 2012. http://phys.org/news/2012-07-seagulls-benefits-climate.html

Seguin, Maria Susana. *Science et religion dans la pensée française du XVIIIe siècle: le mythe du déluge universel*. Paris: Honoré Champion, 2001.

Sekula, Allan. *Fish Story*. Rotterdam: Witte de With Center for Contemporary Art and Richter Verlag, 1995.

Semper, Gottfried. *Style in the Technical and Tectonic Arts, or, Practical Aesthetics*. Los Angeles: Getty, 2004.

Serres, Michel. *The Natural Contract*, trans. Elizabeth MacArthur and William Paulson. Ann Arbor, MI: University of Michigan Press, 1995.

Seznec, Jean. *The Survival of the Pagan Gods: The Mythological Tradition and Its Place in Renaissance Humanism and Art*, trans. Barbara F. Sessions. New York: Pantheon Books, 1953.

Sharp, Katie. "Six Eerie Ways That Radiohead Predicted the Future." *Mic*, 9 April 2015. http://mic.com/articles/114190/6-ways-radiohead-foresaw-the-future

Shelley, Mary. *The Last Man*, ed. Morton D. Paley. Oxford: Oxford University Press, 1994.

Shelley, Percy Bysshe. *Prometheus Unbound: A Lyrical Drama in Four Acts, with Other Poems*. London: C and J Ollier, 1820.

Shepard, Francis P. *The Earth Beneath the Sea*. Baltimore: The Johns Hopkins University Press, 1959.

Shubin, Neil. *Your Inner Fish: The Amazing Discovery of Our 375 Million Year Old Ancestor*. New York: Random House, 2009.

Sloboda, Stacey. "Displaying Materials: Porcelain and Natural History in the Duchess of Portland's Museum." *Eighteenth-Century Studies* 43.4 (2010): 455–72.

Sloterdijk, Peter. *Bubbles: Spheres 1*. Los Angeles: Semiotext(e), 2011.

——. *In the World Interior of Capital: Towards a Philosophical Theory of Globalization*, trans. Wieland Hoban. Cambridge, UK: Polity, 2013.

Smith, Kristianna. "Beirut: A Jet-Setter Settles Down." *NPR.org*, 10 September 2011. http://www.npr.org/2011/09/10/140318038/beirut-a-jet-setter-settles-down.

Smyth, Jamie and Kiran Stacey. "Adani Pushes Ahead with Australia Coal Mine Despite Protests." *Financial Times*, 7 June 2017, 17.

Snapes, Laura. "Video: *Beirut* 'The Rip Tide.'" *Pitchfork*, 22 June 2012. http://pitchfork.com/news/46943-video-beirut-the-rip-tide

Sontag, Susan. *On Photography*. Auckland: Penguin, 1982.

Sorenson, Chris. "Ottawa Shows off Bell recovered from the Erebus." *Maclean's*, 6 November 2014. http://www.macleans.ca/news/canada/ottawa-shows-off-bell-recovered-from-hms-erebus/.

Southey, Robert. *The Collected Letters of Robert Southey Part One: 1791–1797*, eds. Lynda Pratt, Tim Fulford and Ian Packer. Romantic Circles Electronic Edition. Appendix 1, Letter A. http://www.rc.umd.edu/editions/southey_letters/letterEEd.26.appendix1.html.

Spary, E. C. "Rococo Readings of the Book of Nature." *Books and the Sciences in History*, eds. Marina Frasca-Spada and Nick Jardine. Cambridge, UK: Cambridge University Press, 2000. 255–275.

"Species Profile and Threats Database." Australian Government Department of the Environment and Energy. Accessed 6 August, 2018. http://www.environment.gov.au/cgi-bin/sprat/public/publicthreatenedlist.pl?wanted=flora.

Sprawson, Charles. *Haunts of the Black Masseur: The Swimmer as Hero*. London: Vintage, 1992.

Stafford, Barbara Maria. "The Eighteenth Century: Towards an Interdisciplinary Model." *The Art Bulletin* 70.1 (1988): 6–24.

Stampfle, Felice. "A Design for a Garden Grotto by Jacques de Gheyn II." *Master Drawings* 3.4 (Winter 1965): 381–83, 443.

Steinbeck, John. *The Log from the Sea of Cortez*. London: Penguin, 2000.

Stengers, Isabelle. "The Doctor and the Charlatan." *Cultural Studies Review* 9.2 (2013): 11–36.

Strong, Roy. *The Renaissance Garden in England*. London: Thames and Hudson, 1979.

Summerscale, Kate. *The Suspicions of Mr Whicher, or The Murder at Road Hill House*. London: Bloomsbury, 2009.

Susik, Abigail. "Surrealism and Jules Verne: Depth of Subtext in a Collage by Max Ernst." *Surrealism, Science Fiction and Comics*, ed. Gavin Parkinson. Liverpool: Liverpool University Press, 2015. 16–39.

Swan, Claudia. *Art, Science, and Witchcraft in Early Modern Holland: Jacques de Gheyn II (1565–1629)*. Cambridge, UK: Cambridge University Press, 2005.

Szerszynski, Bronislaw. "Colouring Climates: Imagining a Geoengineered World." *Routledge Handbook of the Environmental Humanities*, eds. Ursula Heise, Jon Christensen, and Michelle Niemann. London: Routledge, 2017. 82–90.

Taylor, Bernard. *Cruelly Murdered: Constance Kent and the Killing at Road Hill House*. London: Souvenir, 1979.

Tenison-Woods, J. E. "A Trip to a Coral Reef." *The Australasian*, 18 October 1879, 6.

Thacker, Eugene. "Black on Black." *The Public Domain Review, Selected Essays, Vol. III*, ed. Adam Green. Cambridge, UK: Public Domain Review, 2016. 13–29.

"Thistlegorm Dive Site." Accessed May 2018. http://www.eazydivers.com/eng/index.php/thistlegorm.

"The Thistlegorm Project." Accessed May 2018. http://thethistlegormproject.com/.

Thomas, Keith. *Man and the Natural World: Changing Attitudes in England 1500-1800*. London: Penguin, 1983.

Thompson, John. *Hurley's Australia: Myth, Dream, Reality*. Canberra: National Library of Australia, 1999.

Torre, Stephen. "Tropical Island Imaginary." *Etropic: Tropics of the Imagination 2013 Proceedings* 12.2 (2013): 246–73.

Twist, Stanly H. Foreword to Jules Verne, *Twenty Thousand Leagues Under the Sea*. New York: Grosset and Dunlap, 1917. v–xi.

——. "An Introduction to Umwelt," trans. Gosta Brunow. *Semiotica* 134 (2001): 107–10.

Uexküll, Jakob Von. *Theoretical Biology*, trans. D.L. Mackinnon. New York: Harcourt, Brace, 1926.

"Under-Sea Photography." *The Argus* [Melbourne], 18 February 1916, 5.

United Nations Department of Economic and Social Affairs, Population Division. *World Urbanization Prospects: The 2014 Revision, Highlights* New York: United Nations, 2014. https://esa.un.org/unpd/wup/publications/files/wup2014-highlights.pdf.

University of Arizona. "First evidence of surprising ocean warming around Galápagos corals." *ScienceDaily*. Accessed May 2018. https://www.sciencedaily.com/releases/2018/02/180221131836.htm.

Valéry, Paul. *The Outlook for Intelligence*, trans. Denise Folliot and Jackson Mathews. New York: Harper and Row, 1963.

Van Regteren Altena, I.Q. *Jacques de Gheyn: Three Generations*. 3 vols. The Hague: Martinus Nijhoff Publishers, 1983.

Van Reijen, Willem. "Breathing the Aura – The Holy, the Sober Breath," trans. Claudia Kotte. *Theory, Culture & Society* 18.6 (2001): 31–50.

Verne, Jules. *Twenty Thousand Leagues under the Sea*, ed. and annotated by Walter Miller. New York: Thomas J. Crowell, 1976.

Vieiri, Francesco de'. *Discorsi di M. Francesco de' Vieri, detto il Verino secondo, cittadino fiorentino, delle marauigliose opere di Pratolino, & d'amore*. Florence: Appresso Giorgio Marescotti, 1586.

Vince, J. and B. D. Hardesty "Plastic Pollution Challenges in Marine and Coastal Environments: From Local to Global Governance." *Restor. Ecol.* 25, (2017): 123–28.

"Virtual Tour of Second World War Shipwreck Goes Online," *The Guardian*, 6 October 2017. https://www.theguardian.com/world/2017/oct/06/virtual-tour-of-second-world-war-shipwreck-thistlegorm-goes-online.

Von Mallinckrodt, Rebekka. "Exploring Underwater Worlds: Diving in the Late Seventeenth-/Early Eighteenth-Century British Empire." *Empire of the Senses: Sensory Practices of Colonialism in Early America*, eds. Daniela Hacke and Paul Musselwhite. Leiden and Boston: Brill, 2018. 300–22.

Walker, Gabrielle. *An Ocean of Air: A Natural History of the Atmosphere*. London: Bloomsbury, 2010.

Wake, Bronwyn. "Snapshot: Snow White Coral." *Nature Climate Change* 6 (2016): 439.

Warner, Marina. *From the Beast to the Blonde: On Fairy Tales and their Tellers*. New York: Farrar, Straus and Giroux, 1994.

Watson, William. "An Account of a manuscript treatise, presented to the Royal Society, intituled *Traité du corail*." *Philosophical Transactions* 47 (1751–2): 445–69.

Wellek, René. "The Concept of Baroque in Literary Scholarship." *The Journal of Aesthetics and Art Criticism* 5.2 (Dec. 1946): 77–109.

The West Australian, 1 September 1893.

West, Michael. "Spenser, Everard Digby, and the Renaissance Art of Swimming." *Renaissance Quarterly* 26.1 (1983): 11–22.

Westling, Louise. *The Logos of the Living World: Merleau-Ponty, Animals, and Language*. New York: Fordham University Press, 2014.

——. "The Anthropocene and the Long 17th Century 1550–1750," *The Cultural History of Climate Change*, eds. T. Bristow and T. Ford. London and New York: Routledge, 2016. 87–107.

Williams, John W. et al. "Model Systems for a No-Analog Future: Species Associations and Climates during the Last Deglaciation." *Annals of the New York Academy of Sciences* 1297 (2013): 29–43.

Williams, Linda. "Between Hermes, Gaia and Apollo 8: Michel Serres and the Philosophy of Science as Communication." *Access: Critical Perspectives on Communication, Culture and Policy Studies* 26.2 (December 2007): 33–45.

——. "Darwin and Derrida on Human and Animal Emotions: The Question of Shame as a Measure of Ontological Difference." *New Formations: A Journal of Theory/Culture/Politics* 76 (Summer 2012): 21–37.

——. "Reconfiguring Place: Art and the Global Imaginary." *The SAGE Handbook of Globalization*, eds. Manfred Steger, Paul Battersby, and Joseph Siracusa. London: Sage Publications, 2014. 463–80.

Zipes, Jack. *Fairy Tales and the Art of Subversion: The Classical Genre for Children and the Process of Civilization*. New York: Taylor and Francis, 2006.

——. *Hans Christian Andersen: The Misunderstood Storyteller*. New York: Routledge, 2005.

Index

Page numbers in italics refer to figures. Page numbers followed by 'n' refer to notes.

9780367001582

9780367001582